Becoming Native to This Place

THE BLAZER LECTURES FOR 1991

BECOMING NATIVE *to* THIS PLACE

WES JACKSON

THE UNIVERSITY PRESS OF KENTUCKY

Library of Congress Cataloging-in-Publication Data

Jackson, Wes.
    Becoming native to this place / Wes Jackson.
        p.   cm.—(Blazer lectures ; 1991)
    Includes bibliographical references.
    ISBN 0-8131-1846-8 :
    1. Environmental responsibility—United States.   2. Environmental
policy—United States.   I. Title.   II. Series.
GE195.7.J33   1994
363.7'00973—dc20   93-41335

For Wendell Berry

Contents 🦁

Foreword 🦁

The necessity for social progress—economic, political, cultural, technological—has long been such a bedrock assumption in the American ethos that challenges to it have rarely been taken seriously. True, it is admitted, there are certain costs attendant upon making progress, but these costs are seen as slight compared to the benefits. Moreover, the usual assumption is that many of the costs derive from lamentable—but in principle avoidable—excesses. If early twentieth-century miners despoiled the fragile richness of the Rockies, the fault lay not with "progress" (and its concomitant voracious search for industrial raw materials) but rather with an earlier generation's lack of consciousness about environmental impacts.

Still, there have been other voices. Norman Ware, the most eloquent chronicler of the demise of the American craftsman under the pressure of industrialization, put it this way:

It is commonly supposed that the dissatisfaction in the [1840s] with the character and results of the Industrial Revolution was the result of purely temporary maladjustments. It is admitted that a temporary maladjustment lasting over one's working lifetime is sufficiently permanent for the one concerned. But it is claimed that, from the standpoint of history, the degradation suffered by the industrial worker in the early years of the Industrial Revolution can be discounted by his later prosperity. And this might be true from the calm standpoint of history if the losses and gains were of the same sort. But they were not. The losses of the industrial worker in the first half of the century were not comfort losses solely, but losses, as he conceived of it, of status and independence. And no comfort gains could cancel this debt. [*The Industrial Worker, 1840-1860* (Boston: Houghton Mifflin, 1924), pp. x-xi.]

Wes Jackson now adds his powerful voice to this cautionary literature. For Jackson, too, the losses from unbridled and thoughtless economic "growth" are not simply comfort losses— more fast food paid for by less access to parks—but rather the disruption and destruction of some of the most fundamental conditions for leading a fulfilled and meaningful life. His concern focuses on modern society's tendency to deprive us of our connections to our natural and social environment. The contrary movement, a conscious choice of will, is to become "native to this place."

Mr. Jackson, founder and director of The Land Institute near Salina, Kansas, delivered the 1991 Blazer Lecture at the University of Kentucky. This volume is an extension of his talk. The Blazer Lecture Series, made possible by generous grants from the Paul G. and Georgia M. Blazer Fund and the Blazer family, have long enriched campus life, and we are grateful for their support.

Richard Edwards
Dean of Arts and Sciences
The University of Kentucky

Acknowledgments

Nothing more underscores the reality of the near endless web of interactions among my friends and colleagues than when the time comes to list my intellectual debts. That debt load is now huge, and the intellectual interest compounds more than semiannually. I have no choice, therefore, but to hereby declare all such debts as gifts.

The gift of deepest insight into most of the issues and proposed solutions discussed here came from my friend Wendell Berry. As a small statement of appreciation, I dedicate this book to him.

There are other friends who have been of immense importance for their intellectual contribution, their living example, or both, including: Tanya Berry, David Ehrenfeld, Terry Evans, Rick and Joyce Fent, Leland Lorenzen, Steve Marglin, Gary Nabhan, Conn Nugent, David Orr, J. Stan Rowe, Arnold Schultz, Charlie Sing, Douglas Sloan, John Todd, Charles Washburn, Don Worster, Angus Wright, and Arthur Zajonc.

Colleagues at The Land Institute Jon Piper, Peter Kulakow, and Marty Bender, one way or another before and during the preparation of this effort have helped give shape to these ideas. Eight to ten interns come through The Land Institute each year to spend forty-three weeks in reading, thinking, discussing, and doing the necessary physical work around the place. These smart, idealistic college graduates are indispensable in the energy and inspiration as well as the thinking they provide. They are at once a barometer of where the culture has gone and a source of ideas as to where it should go.

The core of the book was the 1991 Blazer Lecture at the University of Kentucky. Sam Hileman made numerous valuable improvements. Linda Okeson at The Land Institute typed various versions of the manuscript more times than the word processor can remember, and now and then would say, "Do you want to say this?"

The Pew Charitable Trust, through its Conservation Scholars program, provided funding that has made it possible to begin implementation of some of these ideas in a small Kansas town. By being able to enact these ideals in a real physical place, more ideas have emerged which have informed much of the content of this book.

Prologue 🐾

When one of my great grand-fathers swept into Kansas with the white tide on May 30, 1854, the first day he and the others could legally do so, the day the Kansas-Nebraska Act was signed by Franklin Pierce, our nation had fewer than 30 million people. Had national policy at that time been directed toward urging all Americans to become "native" to this place, the nature of our relationship to the land today would be very different from what it is. Today, too many people and the products of the technology explosion, interacting with our desires and our perceived (as well as bona fide) needs, dictate the terms of that relationship.

It was always changing. By the time one of my grand-fathers (the above-mentioned great grandfather's son-in-law) made it to Kansas from the Shenandoah in 1877, the standard we might have employed for an 1854 "nativeness" was already rapidly disappearing. The great herd of bison was nearly finished off. The Santa Fe Trail, at age fifty-six, as an *official* highway of commerce, would soon become totally irrelevant. And by the time that grandfather died in 1925, 45 million acres of pristine prairie had been broken by tractors and horses and planted to wheat. I was born eleven years later, at the height of the Dust Bowl era, which was a consequence of that Great Plowing. It was an era in which the heart of our continent sent its finest soil particles far overhead to Washington and even to ships at sea.

It has never been our national goal to become native to this place. It has never seemed necessary even to begin such a journey. And now, almost too late, we perceive its necessity. Unfortunately, the nature of the nativeness toward which we must work has been not merely altered but severely compromised. Part of the reason is that we have eight and a half times as many people in our country as we did when my grandfather was born. Perhaps even worse, the forces that have given us our modern problems—the ozone hole and global warming, acid rain, Three Mile Island and Chernobyl, soil erosion and loss of family farms, and so on—gain power by the decade. Destruction is occurring at an accelerating pace. It has all happened so fast (more than 80 percent of all the oil ever burned has been burned in my lifetime) and it is going to get worse— half of Mexico's population is under fifteen years of age, ready for a major explosion. The world is slated to add one billion people in the 1990s alone. More people will be added in ten years than the total population of the earth at the time of Columbus.

This book is dedicated to the idea that the majority of solutions to both global and local problems must take place at the level of the expanded tribe, what civilization calls community. In effect, we will be *required* to become native to our little *places* if we are to become native to this *place*, this continent. Although we have told one another on bumper stickers and at environmental conferences that we must "think globally and act locally," we tend to drift toward mega-solutions. Rather than get busy, we introduce new terms such as "sustainable" to apply to any perceived solution that catches our fancy. Instead of looking to community, we look to public policy. We hold a global conference in Rio.

To a large extent, this book is a challenge to the universities to stop and think what they are doing with the young men and women they are supposed to be preparing for the future. The universities now offer only one serious major: upward mobility. Little attention is paid to educating the young to return home, or to go some other place, and dig in. There is no such thing as a "homecoming" major. But what if the universities were to ask seriously what it would mean to have as our national goal becoming native in this place, this continent? We are unlikely to achieve anything close to sustainability in any area unless we work for the broader goal of becoming native in the modern world, and that means becoming native to our places in a coherent community that is in turn embedded in the ecological realities of its surrounding landscape.

This is not just another way of talking about sustainability or bioregionalism, though both would be the result to some degree. The subject is broader than that, for it is largely cultural and ecological in scope.

The first natives here were not burdened with the exercise of technology assessment, as we must be. The technological array developed during the industrial revolution needs to be inventoried and assessed for a world approaching the end of fossil fuel.[1] That won't be easy. As we begin to anticipate what we can or should take with us into that sun-powered future, we will soon discover how limited our imaginations are. Quite frankly, one of my major worries is that we will become so discouraged that we'll seek to repeal Murphy's law and opt for nuclear power. (We will do this all the while explaining to ourselves that it is necessary if we are to avoid widespread social upheaval.) There are bound to be numerous surprises once we get into the inventory and assessment stage. The global market has given us so many intersecting loops and our economic ecosystem is so complex that the fanciest systems programs cannot accurately predict what we will be able to keep and what will be selected against. This alone is argument enough for that second major, the "homecoming major." I am not talking here about mere nostalgia. To resettle the countryside is a practical necessity for everyone, including people who continue to live in cities. To gather dispersed sunlight in the form of chemical energy in a fossil-free world will require a sufficiency of people spread across our broad landscape, even though we may be more efficient in a sun-powered future than in the past by employing some modern technological equipment. The area over which we will have to range to collect the sunlight may be so large that the economics that would follow the energy cost may make it prohibitive.

This resettlement will be no small matter. It will have to be carried out by those who have a pioneering spirit, by those who see the necessity of such a dispersal, by those intelligent enough and knowledgeable enough about its necessity that they will have the staying power. What they will be up against is horrendously formidable: a society dominated by the rich and powerful, offering temptations to embrace the extractive economy that keeps our incomes and the global nonrenewable resources flowing their way.

Think of what such a shift would mean for our universities. Today they hold the majority of our young people hostage for four years with the always implicit and often explicit promise of upward mobility. For tens of thousands of students, the universities have become little more than holding pens that keep them off the job market, requiring them to devote millions of hours to turning out work too shoddy to be either useful or artistic. Think about what is likely to be the eternal judgment of the generation now in power. As the result of its excesses, this generation is likely to be the first, and for that matter the last, after it has died off, at best to be regarded as simply comical and pathetic and at worst to be hated. Isn't it time we begin figuring out a way to earn a living and amuse ourselves cheaply, which is to say with the least expense to our life support system? The binge the developed world has enjoyed is about over. It's time to find our way home and use what little time is left for partial redemption of this prodigal generation.

1 The Problem 🐾

In 1992, the people of the Americas acknowledged and celebrated Spain's *entrada* into the New World half a millennium ago. Few remembered that half a century after that event a young crew of Spanish adventurers were dispatched into the heart of the North American continent to locate the mythical Seven Cities of Cíbola. They penetrated the continent to what is now central Kansas. The trek of these young conquerors amounted to the establishment of a line that would divide history and prehistory.

The Coronado expedition of 1540-1542 began when Francisco Vázquez de Coronado left Compostela, Mexico, and headed north toward Culiacán with his troop of 336 Spanish soldiers, plus the wives and children of a few of those soldiers and several hundred Indians. The march to Culiacán was a preparatory shakedown. When the army left Culiacán, 250 of the Spaniards were on horseback, and more than a thousand horses and mules were packed with baggage, arms, provisions, and munitions. There were six companies of cavalry, one of artillery, and one of infantry, and several friars who walked. The expedition marched off northward in February of 1540.

A year and a half later—in June of 1541—Coronado and about thirty of his men reached the Indian kingdom of Quivira. By early August they achieved the northernmost point of their march, the northeasternmost village of Quivira in what is now central Kansas. Coronado had hand-selected this smaller group of men for this side trip. They were mostly in their twenties; Coronado himself was only thirty-one. All were irritated that they had not found the rumored wealth they sought when they finally arrived at Cíbola, now the Zuni Reservation in Arizona. They had been enticed off on this second wild goose chase into what is now Kansas by the lies of an Indian slave who wanted to get home to his people, the Harahey, a people who resided in either northeastern Kansas or southeastern Nebraska. The Spaniards called this man the Turk. He had deceived them with one major lie and buttressed it with a series of lies concerning the great wealth of the kingdom of Quivira. Quivira had plenty of people and good land. They were tall people; one stood six feet eight inches. But the houses of Quivira were built of grass and sticks. There was no gold. Even Chief Tatarrax wore only a copper ring on his neck.

Coronado finally, reluctantly gave in to the pressure of his angry subordinates. They were allowed to strangle the Turk. Thus young noblemen from some of Europe's finest families were responsible for the first murder of an Indian by whites in Kansas.

Frustrated and out of sorts, the Spaniards turned back. The kingdom of Quivira seemed too small and poor to deserve their conquest. Something behind their European eyes prevented their seeing what was before them. The Quivirians were skilled workers in stone. Their pipes, many carved from Minnesota pipestone, were graceful and finely polished. They were great traders; worked stone not native to their location was present. They were armed with bows and arrows, lances, spears, clubs, tomahawks and slings. The best bows were made of horn. Arrows were made of dogwood and hickory.

On the basis of evidence from dwellings and villages that have been excavated, and that from written narratives by those who accompanied Coronado or came into the region later, archaeologists have estimated that "within the confines of Rice County [Kansas] there were well over 25,000 people,"[1] or about thirty-five natives per square mile. In 1927, Rice County had just under 15,000 people. In 1980, 11,800. In 1988, 10,800. In 1990, 10,400.

Why this huge decline in numbers of people? Were the natives more sophisticated at providing their living than we are? More than 25,000 people are now being supported in cities outside the county by Rice County soils and water, by steel produced in Gary, Indiana, and by fossil fuel. We know that nearly all of the young people who want to stay or who leave but want to return would bring the population to well over 25,000. Why can't they stay or afford to return?

A neighbor of mine, Nick Fent, has recently written up the seventy-year history of 240 acres (three contiguous eighties) fifty to sixty miles north of old Quivira. Nick and his wife Joyce now own those 240 acres. The topography, the rainfall, and the soils are comparable to those in old Quivira. Nick

Fent used courthouse deedbooks in his research. The three
eighty-acre tracts were originally deeded to three settlers who
had come west to establish new lives for themselves and their
families. Nick Fent describes how "buried in the legal abstracts
is the depressing struggle of these transplanted families who
had bet mortgages against drouths, dust storms and grasshop-
pers. The documents recording the satisfaction of mortgages
are always preceded by those recording another mortgager of
larger debt." In 1891, a woman living on this land went insane,
and her husband, "owing to [his] extreme poverty," could not
transport his wife to the asylum at Topeka. Nick Fent continues:

Subsistence farming on the center 80 acres had been subjected to
drouth, when crops failed completely; wet years, when the creek
bottom corn fields were too soft to work; and plagues of grasshop-
pers that ate everything from cornstalks to hoe handles. Burdened
with increasing debt, it had passed through 14 owners from the time
of its presidential patent deed No. 1, in 1885, to our ownership in
1955. Through 70 years, owners Dell, Hawkins, Decious, Crowel,
Minor, Carnal, Curtis, Loughridge, Scholl, Haley, Ashman, Wetchel,
Walker and Mills tried, through their struggle with the elements,
insanity, death and taxes, to eat, educate their children and pay the
interest on their debts.

    Most of their personal struggles and disappointments were
hidden in the quiet desperation of their lives but are reflected in the
permanent record of escalating mortgages. One can still see a few
piles of disintegrating horse harness in the corners of dying barns;
old sandstone block building foundations; rock walled hand dug
wells and cisterns and small rock quarries on the hillsides. Faint farming furrows can still be traced through now-forested small creek bottom fields. Old trails, cutting diagonally through the upland pastures, predate the rectangular road grids and show distinctly in the

winter, when snow blows across the prairie to settle in the ancient ruts in white scars across the land.

Life on the adjoining two eighties was not very different. Nick concludes with this chilling sentence: "The families who devoted their lives to losing this land might have prospered elsewhere had it remained 'Buffalo Range.'"

Why did those families fail where the natives had been successful?

Harry Mason, an eighty-three-year-old friend, a former professor of psychology, describes weekend visits from college to the family farm at the county-seat town of WaKeeney, Kansas, in 1934. He tells how the dust drifted under the shingles and eaves to be deposited on the top of ceilings, which then fell through. The plaster and the dust were carted away to fill in a pair of remnant dugouts of failed settlers in an abandoned hog pen.

Harry Mason's father had helped establish the industrial revolution on this near-last frontier of the Great Plains, beginning in the 1890s. His father did custom work with his threshing machine. He farmed, went broke, started a garage business in WaKeeney, succeeded, and lived to witness the evolution and improvement of both tractors and automobiles. But more bad times came. The industrial revolution had every chance here, for it was helped along by the boosterism described by Sinclair Lewis. Here on the 100th meridian, halfway almost to the mile between Kansas City and Denver, two brothers ran the butcher shop and locker plant they had taken over from their father, who had come to the area under contract to butcher the bison that fed the workers building the Union Pacific. From bison to locker plants in less than one generation on the Great American Frontier! And what became of it all?

Harry came home for his father's funeral in 1944. His father had been found dead in his pickup. He had been drinking, a man who in his youth would take no more than a social drink, a tough old man who had refused to leave during the mass exodus of people and blowing dust.

In *The Unsettling of America* Wendell Berry has written that "we came with vision but not with sight. We came with visions of former places but not the sight to see where we are." Out here, in mid-continent, our ancestors arrived with their humid area mindsets, relying heavily on subsidies from the distant East for a startup. But distant subsidies may not be the crucial distinction. Those 25,000 Quivirians whose dwellings were inside what today is Rice County, Kansas, did not do all of their shopping locally either. They traded from Minnesota to Mexico City. They relied on millions of bison calories each year, brought across the current county boundary to their grass huts, calories stored from short grasses grazed from New Mexico to the Canadian prairies. The hides and the horns and the shoulder blade scrapers came from the Great Plains commons.

Nutrients harvested by the bison over the prairie land that became WaKeeney's courthouse square were transported by solar-powered bison legs until they were within reach of those flint-tipped ironwood arrows. Set against the rawhide string of a bow bent almost to the breaking point, those arrows would penetrate almost to the feather. Here was the combination of land and cultural artifact that served as the source of the bones and sinews of Quivira's growing children. Here were the nutrients and here was the stored sunlight that sponsored the countless dives of naked young brown bodies into the deep holes of the Smoky Hill River. This is what sponsored the giggles and excited talk about the freshwater clam brought from the bottom and held high overhead, the wild-eyed excitement of having felt near the bottom of the dive what must have been a ten-pound channel cat slip by. The bison mowed down the green molecular traps set by such star-grasses as blue grama, hairy grama, big bluestem, Indian, and switch grass, unwittingly storing it for these children while chewing their cud as they nooned in the cottonwood's shade.

And so I return to my question: Why has our culture, which insists that we plant wheat where the grass huts of old Quivira stood, failed so miserably at finding ways to support as large a population on this land as the natives did? Why did fourteen families fail in seventy years on one piece of ground? Why do we still lose people when the rate of energy use and the rate of nutrient flow is at an all-time high? We have *sent* our topsoil, our fossil water, our oil, our gas, our coal, and our children into that black hole called the economy.

But I have already warned against a simple indictment of long-distance movement of materials and energy. The problem has to do with the nature of the outside subsidy. Those subsidies from outside Rice County that sustained the Quivirians were of a different order from what Rice County residents receive today. The calories stored in the meat and hides of the buffalo represented contemporary sunlight, not the ancient fossil variety. The bone and other materials that Quivirians used to create tools and clothing represented an acceptable use of nature. The age of the energy package they broke open, be it bison brisket or the sticks to cook it, were measured in tens of years or less. Most of the energy we use today is ancient; our fossil fuel comes from energy packages hundreds of millions of years old; the electricity from our nuclear power plants comes from packages of energy billions of years old—nearly as old as the universe.

With all of this ancient energy available, why are there 10,400 people in Rice County today? Why a *consolidated* Quivira Heights High School instead of Bushton High and Geneseo High and Holyrood High and Lorraine High? Why is it that so many young white people who love the land of Quivira can't make a living there?

2 Visions and Assumptions 🐾

$\mathbf{W}$endell Berry's classic *The Un-settling of America* describes the sequence of conquest and settlement. Natives, not "redskins," were living on this land to which European conquerors came. From the moment these natives became "redskins," they became surplus people; the "redskin" designation validated killing them off or moving them off, making their land available for *our* settlement. Without realizing it, we established a precedent. In due time the descendants of those settlers also became surplus people—the new redskins, so to speak. The old farm families were removed and their rural communities destroyed as the industrial revolution infiltrated agriculture.

Just as the natives who became surplus could have shown us how to live harmoniously on the land, even with some of our European cultural modifications, so the surplus farmers now gone could have passed on their myriad cultural techniques, some developed here, others adapted from our agricultural origins in Europe. They never really had a chance. They were moved too abruptly off the farm, out of the small towns, into the cities.

The conqueror is nearly always from someplace else, as Wendell Berry says. In the old days he came as a seeker of gold or markets and sometimes as a mere pawn in European power politics. In the last round of conquest here, the market seeker came bringing the machines and chemicals and the agricultural economists who said "get big or get out."

Conquerors are seldom interested in a thoroughgoing discovery of where they really are. Three days after Columbus arrived in the New World, he wrote in his journal, "These islands are very green and fertile and the breezes are very soft, and it is possible that there are in them many things, of which I do not know, because I did not wish to delay in finding gold." Six days later he wrote, "The singing of little birds is such that it seems that a man could never wish to leave this place." But this man had a mission, and so he left. Missions of conquest seldom have much to do with natural "greenness," "fertility," "soft breezes," or "little singing birds."[1] The man was looking for gold!

Fewer than fifty years after Columbus, Coronado was looking for gold, too. Like Columbus, he could not help but notice the countryside and comment on how handsome and bountiful it was—though merely as a side attraction. When the futility of his quest for gold in Quivira became apparent, he and his small crew of young noblemen turned their backs on the productive landscape that had touched a deeper human sense in them.

And therein lies the tragedy. We are still more the cultural descendants of Columbus and Coronado than we are of the natives we replaced. Now that we find ourselves in a cycle of transition from conquerors of "redskins" to settlers to sons and daughters of settlers who have become the new "redskins," we realize we must break the cycle. But even that will not be enough. This time, to become native to this place we will have to take measures to reduce the chance of ever becoming "redskins" again.

Professor Dan Luten, retired from the University of California Geography Department, has written that we came as poor people to what we perceived to be an empty land rich in resources; now we have become rich people in an increasingly poor land that is filling up.[2] Our institutions were built on a former reality and don't do well in the modern context. A sobering question is: Does an experiment such as the one we Americans have wrought work only where there is the kind of slack that this yet unused-up continent once afforded?

On August 24, 1874, a party of six General Land Office surveyors led by a Captain Short were attacked eight miles southwest of Meade, Kansas, as they were laying out township section lines. All six were killed and three were scalped. This was called the Lone Tree Massacre because of a lone cottonwood that stood on the spot until blown down in 1938. We don't know the motivation for the massacre. But the surveyor's instrument was symbolic of the difference between the two races.

Earlier I mentioned WaKeeney's courthouse square, how the bison must have freely roamed over that small piece of land, nibbling on the buffalo grass that helped support the lives of Quivira's children. Not far away stands a similar courthouse where the patent deed on the Fent eighty acres was processed. The township section lines became the basis for land distribution, including the Fent land where fourteen families failed over a seventy-year period. The Indians had no such lines. Nutrients and sunlight picked up by the bison in the neighborhood of the Mandan Indians of North Dakota, for example, were likely harvested by Quivirians in Kansas, and vice versa. "Holding" nature as a commons was a way of spreading the risk. It blunted the extremes of floods and drouths, cold and heat.

In *Wolf Willow*, Wallace Stegner tells of the Canadian minister of Public Works who in 1869 ordered a military colonel to select "the most suitable localities for the survey of townships for immediate settlement." Stegner describes how the rectangular surveys "would cut across the little farms that the *métis* had established on the Assiniboine, the Red, and the other rivers." Before the surveys there were

long strip farms, each with a frontage on the river which gave not only a canoe landing but an access to water for the irrigation of gardens. The strips ran far back and were combined in common pastures. On these pastures the *métis'* stock could run freely while people were off on the annual hunts. The processes of adaptation to Plains life and to the uncertain rainfall had led the *métis* to an economy not unlike that of the Apache after the acquisition of the horse. They were half horticultural, half nomadic, and their system of land division was appropriate to their life. It was far better adapted to the arid and semi-arid Plains than the rectangular surveys were, but nobody in Canada or the United States understood that.[3]

The grid and property lines and what they mean must be factored in, almost as immutable givens, as we begin our journey to become native to the place. Those lines are likely to last as long as there is a United States.

One lesson: Human history forces upon us the terms of our coming nativeness as much as or more than does our freedom to choose. At the time of the 1874 massacre, we still had a chance for a kind of pastoral commons on the Great Plains. Few fences were up. The great bison herd was still intact. Natives were in a steep decline but still around. We had relatively little to keep us from achieving the ecological resilience the Quivirians of Coronado's time had known. But by 1885, eleven years later, the year the first patent deed was issued on the Fent property, a quantum leap had been made. The grid was absorbing settlers within well understood property rights. By 1900, nearly every quarter section was occupied. More options to build a sustainable future were lost between the 1874 Lone Tree Massacre and the end of the century than from before the massacre all the way back to Coronado! Now a different sort of nativeness would be required.

The longer we wait, the more complicated will be our journey to become native. The definition of what it means to become native will go on changing as our options continue to lessen. Temptations to compromise, on the other hand, will increase. More will be required than merely planting and allowing more trees and welcoming back the wild animals. New England today has more trees than it had in the time of Thoreau, but they have multiplied largely because of the shift to heating oil. And though we are regaining some wild animals now through game ranching and wild species reintroduction, we have less and less topsoil, fewer species, less germplasm in our major crops, and more shopping centers.

From our first arrival we have behaved as though nature must be either subdued or ignored. In the early seventeenth century, scarcely a hundred years after initial European arrival in this hemisphere, the Englishman Francis Bacon said that if we are to have the good life, nature's secrets will have to be forced from her. In this he was explicit, for he says that "neither ought a man to make scruple of entering and penetrating into these holes and corners, when the inquisition of truth is his whole object."[4] The pursuit of "truth" justifies anything! Solomon's house in Bacon's utopia, *New Atlantis*, is the prototype for the modern scientific laboratory. (This utopian essay, by the way, served as a model for the Royal Society.) Descartes, in agreement with Bacon, advanced the idea that we grant priority to the parts of things over the whole. Scientists have followed these instructions. More or less ignoring the fact that emergent properties define the various fields of science, they have denied or ignored the profound reality of interpenetration of part and whole—the clear reality of the whole affecting its parts even as those parts affect the whole. This nearly inescapable methodology of science has now been practiced for well over three hundred years. Even the best scientists have believed that the world operates by the same method they use to study it.

In *The Dialectical Biologist*, Richard Levins and Richard Lewontin, Harvard biologists, have outlined the ways in which the Cartesian worldview has created problems for humanity.[5] In arguing for a dialectical view, they suggest that we look again at the relationship between part and whole and acknowledge that part influences whole and whole influences part. This view that the world is made up of stacked dead atoms, ultimately under-standable in some quantitative sense, leads us to think that all life forms—trees, deer, and humans—are "nothing but," a view that has contributed to our alienation from nature.

Observation of activity at all levels of organization dem-onstrates that this is not true. For example, gene splicers have spliced into the bacterium *E. coli* some hereditary material from a human cell. The nucleic acids in sequence make up the code for a specific linear arrangement of amino acids. The amino acid sequence in this chain is presumably the same in the bacterium as in the human cell. So far, so good. Part has priority over whole. But eventually this linear chain begins to bond to itself in various places, forming little loops here, little loops there, so that by the time it is through bonding, there results a three-dimensional structure that in the human cell is capable of some essential biological activity. But in this exam-ple, the sequence of amino acids that is at home in the human cell, when produced inside the bacterial cell, does not fold quite right. Something about the *E. coli* internal environment affects the tertiary structure of the protein and makes it inac-tive. The whole in this case, the *E. coli* cell, affects the part—the newly made protein.[6] Where is the priority of part now?

Another example comes from work done in embryology on mammals.[7] Immediately after fertilization in a normal

mouse, the egg cell contains two separate bodies called pronuclei. One was already present in the egg before fertilization
and contains the female complement of chromosomes. The
other is from the sperm and carries the male complement of
chromosomes. If the male-donated pronucleus is removed and
a female-sponsored pronucleus is inserted, so that now both
pronuclei have their origin in females, development stops.
When the opposite condition is imposed, two male pronuclei
in the egg, the result is the same. The conclusion is that regardless of whether the genetic material is represented in du-
plicate, it won't work unless there is one of each from each sex.
In other words, mammals need moms and dads.

Science is supposed to be value free but the reality is that
our values are able to influence the genotype of our major crops
and livestock. Our minds influence the structure of events. Our
major crops—for example corn, soybeans, and wheat—have
genes that we might call "Chicago Board of Trade genes." There
are also "fossil fuel wellhead genes" and "computer genes." That
is, there are ensembles of genes in our major crops that would
not exist in their particular constellations were there not a Chicago Board of Trade (where a major share of our agricultural
transactions occur), or fossil fuel to make and run farm equipment, or computers to assist agribusiness. The list of genes affected by culture goes on and on. Our *values* affect the arrangement of even the molecules of heredity. *That* is interpenetration.

Gary Nabhan tells a story about a Native American
woman in Mexico who had several ears of corn from her corn

crop arranged before her as she shelled grain from each ear. There were ears that were tiny nubbins and ears that were long; all had seeds of various colors. As she shelled grain from each ear to save for the next planting, Gary asked her why she saved seed for planting from the small ears. Her reply was that corn is a gift of the gods and to discriminate against the small in favor of the large would be to show a lack of appreciation for the gift. What she was doing, in genetic terms, was maintaining genetic diversity. Values dictate genotype. James B. Kendrick at the University of California at Berkeley says that if we had to rely on the genetic resources now available in the United States to minimize genetic vulnerability in the future, we would soon experience significant crop losses that would accelerate as time went by. Roughly one-third of our current crop comes from four inbred lines, which is roughly the same as the amount of variation that may be found in as few as two individual plants.

As the Cartesian world becomes scaled down, what rides with it is the notion that the world is like the method. The dialectical or ecological approach asserts that creating the world is involved in our every act. It is impossible for us to operate in our daily lives and *not* create the world that everyone must live in. What we desire arranges the genetic code in all of our major crops and livestock. We cannot avoid participating in the creation, and it is in agriculture, far and away our largest and most basic artifact, that human culture and the creation totally interpenetrate.

Even though attack on the prioritization of part over

whole is the usual method of bashing Bacon and Descartes, this may not be the most troublesome legacy from this early period of modern science. An even worse bequest came to us in Descartes's *Discourse on Method*, where he says that the more he sought to inform himself, the more he realized how ignorant he was. This statement would have been all right if he had stopped there. Unfortunately, rather than regarding informed ignorance as an apt description of the human condition and the appropriate result of a good education, Descartes believed our ignorance to be correctable. Some ignorance is indeed correctable—by knowledge; but Descartes had an inadequate appreciation of the enormity of his ignorance about the way the world works. The upshot was that we inherited and developed a knowledge-based worldview founded on the assumption that we can accumulate enough knowledge to bend nature pliantly and to run the world. Greenhouse gases are the result of that knowledge-based worldview. So is acid rain; and so is Chernobyl, with its twenty-seven towns and villages evacuated and, the former Soviets say, abandoned forever.

Ozone destruction is a product of that same knowledge-proud, knowledge-confident worldview. The time-line on this process of destruction begins where we fix the beginning of modern chemistry—as far back perhaps as the 1880s, with vulcanization of rubber, the early work on petroleum, and the internal combustion engine; or later still, 1946, when the polymer era began. Whenever its startup was, modern industrial chemistry required less than one century to give us an ozone hole. Like nearly all of us, our chemists have lived in a world so knowl-edge-based that they are scarcely cognizant of their (and our) assumptions. Denis Hayes has noted that around 1970, the time of the first Earth Day, chlorofluorocarbons were on essentially every chemist's short list of good chemicals—nontoxic, non-corrosive, nonpolluting, did a lot of wonderful things. These human-made chemicals were not agents of destruction then, so far as we knew. With this lesson in mind, it is fair to ask how long it will take our biotechnologists to come up with the bio-equivalent of the ozone hole. After all, more money is now going to this new field than went to our chemists in the early years. Just as it was more or less fruitless to argue whether this or that chemical should be released upon nature, it is equally fruit-less to argue whether ice-minus-bacteria should be released, or whether the Beltsville hog with the human growth hormone gene spliced in should exist, or whether we should be using biotechnology to try to develop herbicide resistance in soy-beans. What makes us think we can adequately assess these proposed projects? Only our Cartesian assumptions that we *know enough to run the world*, despite the reality that we are billions of times more ignorant than knowledgeable.

Do we have any alternative?

We could begin by accepting our profound ignorance—that we will never know more than a small part of what we need to know. Beginning with this very different assumption, we are forced to remember our past, to hope for second chances, to keep the scale of our projects small, and to be ready to back out when things go sour.

Here lies my worry. Most proposals for bringing about a sustainable agriculture and culture carry the fingerprints or markings of the Baconian-Cartesian worldview. At best, it amounts to Smart Resource Management. With such a mind-set, and with the institutions in place to accommodate that worldview, attempts to break out of the paradigm will be extremely frustrating. This is what we must do if our thought and labor to become native to this place are ever to bear fruit.

What if we had an ecological worldview as our operating paradigm? An ecological worldview is also an evolutionary view. Time-honored arrangements would inform us of what *has* worked without our running the empirical experiment. Our evolutionary/ecological worldview would inform our decisions, inform our do's and don'ts in scientific investigation. This is another way of saying that we must turn to nature to inform us, to serve as a reference, must turn our thoughts to building a science of ecology that reflects a consultation of nature. Ecology is the most likely discipline to engage in a courtship with agriculture as we anticipate a marriage.

Notice that I say "building a science of ecology." Modern ecology too has suffered from the Baconian-Cartesian world-view. In 1935 the ecologist A.G. Tansley insisted that ecologists isolate the basic units of nature and split up the story into its individual parts. He was trying to move ecology back on track as a rigorous reductionist science, and he was quite successful in doing so.

Few scientists examine the assumptions of their method. Few will engage in a hard critique of Bacon and Descartes. For three years I sat on the technical review committee for my region for the U.S. Department of Agriculture Low Input Sustainable Agriculture grants program. Not only was the science of ecology poorly represented, but both the proposal writers and the evaluators were, by and large, devotees of the Smart Resource Management school and of "solid science" á la Bacon and Descartes.

Since our break with nature came with agriculture, it seems fitting that the healing of culture begin with agriculture, fitting that agriculture take the lead. Earlier I quoted Bacon's dictum that we must "bend nature to our will." Bacon was interested in the good of society overall, but apparently it was agriculture that adopted his clarion call most enthusiastically. Thus the agricultural assumption that nature is to be either subdued or ignored is embedded in a larger cultural assumption. Therefore we should not expect sustainable agriculture to exist safely as a satellite in orbit around an extractive economy.

3 Science and Nature 🐾

It is August 1968, the Tokyo Prince Hotel, the International Congress of Genetics. I am attending my first international meeting. I am a young scientist with a one-year-old Ph.D., pretty full of myself, figuring that my paper, "Introgression and the Maintenance of Karyotypic Integrity," is sure to be a hit. But beyond my little part in the proceedings, I am excited because I know I will see nearly all of the major figures in the world of genetics at the congress, an event held only once every four years. I am excited, not only because it is my first trip to the Orient, or because I will give a paper, or because I will see many famous people in my field, but also because on the program are the names of many geneticists from the Soviet Union. Most are old names known from the literature, scientists the older American geneticists especially will be glad to see because, since the years of Stalin's great purge of Soviet geneticists, the fate of many has been unknown. Some have survived the purge only by turning away from research and publishing on genetics during the last two or three decades, to research and writing concerned more with developmental microbiology or with physiology.

In my very small hotel room at the Tokyo Prince the night of my arrival, after registration, I open the Conference Program and circle the times and rooms where the Soviet papers will be given. I suspect that part of my motivation to attend some of the sessions by these creatures from behind the Iron Curtain is less interest in what they have to say than curiosity about this strange and relatively isolated breed of men. If my memory is correct, more than half the papers by the Soviets were canceled. We learned that, at the last minute, they had been detained, many of them at the airport. Moreover, when one did see the Soviet scientists who were on hand, they were always in a group. I never once saw one alone or talking alone with another scientist. To call it eerie does not adequately describe my feeling.

But one moment at that meeting stood out from all the rest. To appreciate how poignant it was requires some background. The late Theodosius Dobzhansky, one of America's greatest geneticists, was at the congress. He was a truly great man who had written numerous papers and books on genetics, books and papers that have contributed greatly to our modern understanding, especially in the area of population genetics.

Professor Dobzhansky was born in Russia. He had left the Soviet Union before the great purge and had come to America as a young scientist for further training. For three decades while living in New York he heard of the awful and sad events in his home country. Many of those who presumably died in the labor camps were his friends.

Back to the Tokyo Prince Hotel. It was early in the week, the meetings were getting under way, and I was standing in the lobby waiting for an elevator. A small group of Soviet scientists were also standing and waiting. They were all older men. Not one would, I guessed, be under fifty-five or sixty. Some looked to be seventy or older. I managed to catch the eye of a few of them. They nodded to me but quickly turned away, and we all stood in silence, looking up to see where the elevator was on its trip down. It finally arrived. The bell rang. The doors opened and ready to step off was Professor Dobzhansky, face to face and not five feet away from his former countrymen. They stared. He searched their faces. I suspect some of them he had never known, but some he had known were now thirty-five (or more) years older than when, as an aspiring young Soviet scientist himself, he had last seen them. It was a moment when time seemed to stop. Within seconds a wave of emotion swept over the small gathering as one elderly Soviet scientist after another embraced their old colleague. I know nothing of what was said; it was all in Russian, of course. But I do remember, after the bear hugs ended, seeing Professor Dobzhansky put his hand to his right eye, and the flick of his thumb there.

Not wanting to be caught staring, I got on the next elevator going up, aware that I had seen an important moment in the history of genetics. The Soviets had returned and had been welcomed back into the fold of genetics after thirty-five years of self-imposed exile. What led them into that exile has a complex history relevant to our current efforts to develop a sustainable agriculture. It is a history that carries a message I am afraid will be ignored, partly because of human nature and partly because many modern scientists have not taken the time to study some of the subtle implications of the history of science. This is no academic fine point of which I speak. It is far more than that. It has to do with the future of scientific inquiry in agriculture and how that inquiry relates to our future food supply, especially in the developed world. It also has to do with the future of science itself in a world of secular materialism. Eventually, I want to describe what amounts to a complex message from this period of history.

To know for sure where the story begins is hard. We can start with the 1918 Revolution in that part of the world. We can cite some of the political theorists of that time and place. I start with an important Soviet figure, T.D. Lysenko, a man who had a bizarre career. For thirty years, from 1935 until 1965, this fanatical agronomist not only achieved dictatorial power over Soviet genetics, he also wielded power over plant science and agronomy. He used the term "agrobiology" to describe what has been called in the West a "pseudoscientific creation." A man from peasant stock with little formal education, Lysenko had great powers of persuasion. So forceful was he that he persuaded Soviet officials that enormous increases in crop yield could be achieved at little or no cost. Long before his influence came to an end, countless scientists lost their jobs, biological education was almost completely gutted, and classical genetics stopped. According to most historians, his impact on Soviet agriculture was also great and destructive.

To provide a contrast to the worldview and practices of

Lysenko, I turn to another important Soviet of this period, an apparently good guy. The man who stood over and against Lysenko philosophically was Nikolai Vavilov. Born into the nobility of Czarist Russia in 1889, Vavilov was a truly great mind of our century, a geneticist who was internationally recognized for both his theoretical and his practical work as a geneticist. He founded the Institute of Plant Breeding in Leningrad, an institute that played an important role in both Soviet genetics and agricultural research. He was broadly educated, and though he came out of the old nobility, he supported the Revolution and its goals. He had gone to England to study the genetics of wheat. While there, he shifted his interest and placed himself under the influence of the famous geneticist William Bateson, father of the late Gregory Bateson. William Bateson was one of the major interpreters of the father of modern genetics, Gregor Mendel. Mendel's work on inheritance in peas, which founded the school of modern genetics, had been rediscovered at the turn of the century, and William Bateson was there to help shape our modern understanding. Following graduate training under Bateson, Vavilov returned to Russia, committed to the Mendelian-Darwinian worldview. Because he wanted to know the places of origin for our cultivated plants, he became a world traveler with an energy that matched his intellect and passion. His work represented the first systematic attempt to locate the centers of origin of all of our major crops. It was a monumental effort on his part, and thousands of people all over the world assisted him in his collections, as well as in "growing out" tens of thousands of varieties involving the major crops.

At this point, the story takes a tragic turn. Vavilov became the main target of a massive slander campaign headed by Lysenko, who managed to convince Stalin that Vavilov was responsible for the failure of Soviet agriculture following collectivization. After a serious power struggle, Vavilov was finally arrested in 1940. He was brutally interrogated for eleven months and sentenced to death, a sentence he escaped only because he died in jail, probably of malnutrition, in January 1943. Vavilov's former students and numerous supporters were silenced in short order. The Great Purge of Soviet genetics had begun, so sweeping that in 1948 alone approximately three thousand Soviet geneticists stopped all work in their field. Those who survived the purge, some of them the old men I saw waiting for the elevator in the Tokyo Prince Hotel, had shifted fields, as I have mentioned, by working mostly on development and on physiological questions.

Vavilov's work and training were based on the science of the West, science anchored on Gregor Mendel's ideas of independent inheritance, and on Charles Darwin, whose theory of evolution was based on the principle of natural selection.

So far, I have characterized their struggle as one between a good guy and a bad guy in which the bad guy wins for a while but eventually truth wins out. In the main that is correct. But the story was somewhat more complex. Richard Levins and Richard Lewontin, in *The Dialectical Biologist*, point out that in its last years, Lysenkoism was a caricature of the "two camps" view of the world. In the Soviet Union, the confrontation was between the bourgeois camp and the socialist camp. This confrontation within the Soviet Union has been seen as parallel to the confrontation, on the global scale, between the camps of imperialism and socialism. Now that the Soviet Union has come apart, the camp that boasts it has been right all along sends its scientists and social engineers into Central and Eastern Europe to sort things out. Here the interpenetration of history becomes tricky. As Levins and Lewontin point out, the absurdities of Lysenkoism have caused its scientific critics to deny the possibility that both camps may exist within science. They behave, even if they don't say so, as though there is only one true science. In their minds, science really has only one common ground. Moreover, that ground is neutral—that is, it is technically, rationally, indeed totally independent of how it is used. Now that the external political conflict has lessened, this neutral view of science is becoming strengthened at the very moment when the conflict within science needs to be made sharper and recognized as more complex than ever.

Levins and Lewontin succinctly state that the Lysenko movement in the USSR from the 1930s to the 1960s was an attempt at a scientific revolution that failed. They argue that this attempt developed in the following contexts: (1) the pressing needs of Soviet agriculture, which made the society receptive to radical proposals; (2) the survival of both Lamarckian and nonacademic horticultural traditions, the latter of which it drew upon for intellectual content; (3) a social setting of high literacy and a popularization of science, both of which worked to make the debate about genetics a public debate; (4) an incipient cultural revolution, which pitted exuberant Communist youth against an elitist academy; and (5) a belief in the relevance of philosophical and political issues, which put the discussion in the broadest terms. The movement also took place in the context of the geopolitical isolation of the USSR, the Second World War, and the Cold War. Under such conditions came administrative repressiveness and intense philosophical dogmatism. Opportunists jumped on band wagons. The cultural revolution was aborted. Levins and Lewontin conclude:

> In the end the Lysenkoist revolution was a failure, it did not result in a radical breakthrough in agricultural productivity. Far from overthrowing traditional genetics and creating a new science, it cut short the pioneering work of Soviet genetics and set it back a generation. Its own contribution to contemporary biology was negligible. *It failed to establish the case for the necessity of dialectical materialism in natural science.* In the West Lysenkoism was interpreted merely as another example of the self-defeating blindness of communism, but in the Soviet Union and eastern Europe it is still a fresh and painful memory. For Soviet liberals, it was a classic warning of the dangers of bureaucratic and ideological distortions of science, part of their case for an apolitical technocracy.

Buried in this summation is an X-ray of what the revolution in genetics was really about. It was about a dialectical approach to science in opposition to simple reductionism. The Soviets demanded acknowledgment of the reality they perceived in science *and* in society.

Levins and Lewontin have noted that the dialectical view did bring about some breakthroughs in our understanding of the world—the work done on the chemistry of the origin of life, for example. Unfortunately, we are now confronted by the dominance of the idea that in science only one "truth" is possible. Agricultural science in the West suffers today from that "one truth" view. Lysenko and his followers had some nutty ideas about heredity. Had the times been different, we might have long ago modified our methodology of science and the problem of relating scientific method to practical applications in agriculture. Levins and Lewontin make clear that this problem remains with us today as we look to the standard techniques of statistical analysis and the requirement of a control in experiments. I know some modern organic farmers— good ones, too—who challenge the need for such a control, and we conveniently forget that the importance of experimental controls and statistical analaysis was challenged also by the Lysenkoists.

The Soviets wanted heavy doses of philosophy in science. Too many of our scientists assume that philosophy can be and should be excluded from science. But, though some Lysenkoist ideas were absurd, that does not mean that philosophy as such should or can be kept out of science. The philosophical view the Lysenkoists were struggling to foster was more on target than the simple reductionism that still dominates Western science. Our placement of priority on the parts over the whole denies the importance of emergent properties, or qualities, the things that pile up as we go up the level of organization from small to large. Even systems theory is a form of reductionism, for the intersecting variables on the computer do *not* predict emergent qualities. (Take the two gasses hydrogen and oxygen, which combined at a given temperature and pressure give us wetness.) Moreover, when scale is a factor, emergent qualities appear that reflect mere increases in size.

Lysenko was an aberrant, power-hungry nut in a context that brought him to power. Nevertheless, he existed within a philosophical frame we need. For example, the Lysenkoists emphasized traditional, cultural, peasant intelligence as important for agriculture. They made such a fetish of this emphasis that they carried it too far, ultimately suppressing the scientific view held by Vavilov. In our country among our scientists today, it is the other way around, except for the miniature low input sustainable agriculture (LISA) effort and the acknowledged need to develop on-farm research. One could, of course, go overboard in the direction of tradition only. I see it as I travel to various sustainable agriculture gatherings. It is actually a rumble that makes me uneasy. The noise is often from a hard putdown of modern science. When I listen to such putdowns, I remember that in the USSR overemphasis on tradition and cultural wisdom shoved out some good Western science.

The dialectical approach may be the next step toward the "conversation with nature" of which Wendell Berry speaks (see page 40). If we don't acknowledge the interpenetration of part and whole, we are just as likely to drive our landscapes into the kind of ruin the Soviets left of their legacy. Their dialectical science should have been their legacy. A crazy power structure, associated perhaps with the absurdity of some of their ideas about heredity, kept that from happening. Reductionism has a place, but a subordinate one.

⟡ I hope it is now clear that what we think and value shapes the structure of events. This does not imply that the structure of society or the shape of the material world is likely to end up isomorphic with the structure of our consciousness. We are far too clumsy and both society and nature too unyielding to our will. But we cannot help influencing the shape of the world with our thoughts. A major problem is that some of our assumptions about the world and the nature of things lie deep within us, often at the unconscious level. When we attempt to consider other possibilities, these hidden assumptions reassert themselves, usually in unexpected and often undetected ways. Their relative invisibility makes these assumptions resistant not only to change but also to our own criticism. Bad as this is, it may not be our worst problem. More serious is the fact that we have taken too many of our deep-lying assumptions to be "only natural," "just the way things are."

Professor Douglas Sloan of Columbia Teachers College is responsible for articulating the more important thoughts in the above paragraph.[1] In the paper in which he presents them, he goes on to describe five assumptions, all interrelated, that are of special importance in our modern conceptions of knowing and of knowable reality. The first, which he calls "onlooker consciousness," goes back to Descartes, who insisted that we humans are the subject and the only real repository of mind and consciousness. All else is mere object, and mindless. The second assumption is the notion that our senses tell all. Immanuel Kant taught us this. Even the abstractions we conjure up have sense experience standing squarely behind them. Any other possible realities behind or beyond our senses can never be known. The third assumption is that "ultimate reality is quantitative, without consciousness, inanimate"—in other words, raw cause and effect.

With these three assumptions have come the power to analyze, to discover and manipulate the quantitative and mechanistic dimensions of our world. That power has been freeing in a way. The problem is that we have expanded our quantitative view, our mechanistic view, to the point that it has become our all-encompassing picture of the universe as ultimately dead, mechanical, meaningless. This is what is meant when we speak of the scientific and technological worldview, and it has become the background for both the scientific "hard-headed realist" and the technological pragmatist, the utilitarian. This knowledge for its utility, knowledge as instrument for power or control, can be classified as a fourth assumption.

Science didn't start out that way. It was originally a form of inquiry into the truth and nature of things. How it slipped so readily into becoming an instrument to control the world is worth pondering, but not here. For our purposes, it is important to realize that our movement toward strictly utilitarian goals, values, and ends has *narrowed* our rationality by rejecting what Douglas Sloan calls the "earlier truth claims."

Finally, and of equal importance, comes the fifth assumption, which is that, at the most basic level, the universe is acting pretty much the same everywhere. The laws of nature are uniform; things move through space or collide or interact according to universal laws; what we perceive as change has to do only with new combinations and emerging properties of parts. Such a view supports the notion that, since we have succeeded in creating pre-life conditions in a test tube, eventually we will understand all the steps that have led to consciousness. Maybe we will, but so what? Will we still cling to the current idea that human consciousness, once it first appeared, never changed?

By "change" I don't mean that the mind during the experience of history has reinforced some ideas and dropped others. I am not talking about how we can load up with different kinds of ideas, different thoughts that various people have had about the world. I am talking about how, as Sloan puts it, the "quality of consciousness" itself can change, and how this affects what we can experience and know of the world.

We know the benefits of the worldview we are "in" now. We no longer identify ourselves primarily as tribal members alive in nature in a spiritual cosmos. We have instead a great sense of selfhood, with which have come clear-thinking technical consciousness and *control*, leading to capacities for autonomous action. In other words, much of what we call human freedom is a product of our sense of selfhood and our clear-thinking consciousness.

But what about the costs? If this is the only path to all knowledge, it is a path posted with one-way signs directing us toward separation, alienation, abstraction, things quantitative, and all of it at last sense-bound. All of the important qualities we call human—meaning, value, life, consciousness, soul, self, spirit—are off that path of knowing.

So, do we need new ideas, or a new consciousness? Or does it matter which we need? How would we know the difference between a mind generating new ideas and a mind reflecting new consciousness? Maybe this is another subject that is standing in need of further discussion—and that will never know resolution. That may be all right, too, but in the meantime it is something to think about.

An agriculture using nature, including human nature, as its measure would approach the world in the manner of a conversationalist. It would not impose its vision and its demands upon a world that it conceives of as a stockpile of raw material, inert and indifferent to any use that may be made of it. It would not proceed directly or soon to some supposedly ideal state of things. It *would* proceed directly and soon to serious thought about our condition and our predicament. The use of [a] place would necessarily change, and the response of the place to that use would necessarily change the user. The conversation itself would thus assume a kind of creaturely life, binding the place and its inhabitants together, changing and growing to no end, no final accomplishment, that can be conceived or foreseen.

—Wendell Berry[2]

What if researchers in animal science were ever mindful of where our domesticated livestock spent most of their evolutionary history? The chicken would be, first of all (in a breeder's mind), a jungle fowl rather than cackling industrialized property. It would not be just chattel to be confined in small cages to produce eggs or meat. The hog would be a forest animal, not just a meat producer kept in such close confinement that the baby pigs are made into near pin cushions by needles squirting the prophylactic antibiotics close confinement requires. Both the beef and the milk cow had their origin as grazers in savanna-like conditions, not as feedlot critters or milk machines designed to be fed like a hay baler.

None of this suggests an end to science so much as an end to our emphasis on science only as we now know it. The application of our technological array and scientific know-how would become subordinate to a standard emphasizing neither efficiency nor production. Not that we would be uninterested in efficiency or production. As we go about our daily work at The Land Institute, a visitor may not be able to see our motions as different from the motions and patterns of other agricultural researchers. We work in a fancy greenhouse that is under careful temperature control in late fall, all winter long, and into the spring. Here seeds germinate and grow and numerous pollinations are made. We work in orderly experimental plots in the field. But as the breeder pollinates, the pathologist evaluates damage, and the ecologist dopes out soil-root interrelationships, standing firmly in the background is the never-plowed native prairie whose living community is under our close scrutiny.

The products of our effort to make a mental shift can't be readily seen—not yet. I could restrict my argument to life forms. But it isn't just the living world we involve in our mental shift. We acknowledge all of nature. Think about water, for example, and apply the same principle that guides our thinking about how to treat our livestock. Ecological context is foremost: it is one thing for a farmer adjacent to a stream to divert a small quantity of water to irrigate a few acres, and quite another for a stream to become acre feet delivered, via a 400-mile corridor of pipes and ditches, to Los Angeles. Maybe the question is whether the stream, as an ecosystem or part of one, can experience or "enjoy" streamhood. A life form analogy is a few trees removed from a farmer's woodlot to become useful lumber, contrasted with clear-cutting, which converts the woodlot into board feet—a commodity. Acre feet and board feet are the same—resources, an okay notion only if subordinate to streams and forests or woodlots.

The ecological perspective, in other words, honors *jungle* fowl, *forest* animals, and *savanna* grazers—and in the future, I hope, domestic *prairie* seeds. I emphasize the adjectives because each adjective in these cases is the ecosystem that describes the relationship of the larger system to the creature. The ecological perspective honors woods and stream, tropical rain forest, and also Kansas prairie. As we move toward the larger notion of relationship or context the word *resource* loses its significance, for it carries with it the notion of a stockpile for use, even used up if necessary.

I began thinking along this line some fifteen years ago when it became clear that on sloping ground, regardless of terraces and grass waterways, soil would erode on wheat fields, corn fields, soybean fields, sorgham fields—wherever annual monocultures were planted. On the other hand, it was also clear that soil would more or less stay put in perennial pastures, in native prairie grassland, and in forests, independent of human action. But there was more, for to those annual mono-cultures came pesticides, commercial fertilizer, and a need for fossil fuel for traction. The prairie, on the other hand, counted on species diversity and genetic diversity within species, to avoid epidemics of insects and pathogens. The prairie maintains its own fertility, runs on sunlight, and actually accumulates eco-logical capital—accumulates soil. Observing this years ago I formulated a question: Is it possible to build an agriculture based on the prairie as standard or model? I saw a sharp contrast between the major features of the wheat field and the major features of the prairie. The wheat field features annuals in monoculture; the prairie features perennials in polyculture, or mixtures. Because all of our high-yielding crops are annuals or are treated as such, crucial questions must be answered. Can perennialism and high yield go together? If so, can a polycul-ture of perennials outyield a monoculture of perennials? Can such an ecosystem sponsor its own fertility? Is it realistic to think we can manage such complexity adequately to avoid the problem of pests outcompeting us?

At The Land Institute we have confronted these four key questions and have devoted all of our research to answering them. We have turned into action abstractions about nature as measure for agriculture. Ten interns and a staff of thirteen are at work at The Land Institute. To consult nature, we turn to our one hundred acres of never-plowed native prairie; there Dr. Jon Piper, an ecologist, and interns clip above-ground plant material and, after drying, sort it into families of grasses, legumes, sunflowers and "other." By learning how the various ratios of these major plant families differ across soil types and in wet and dry years, and by combining that knowledge with our data derived from soil-root interactions, we then set out to mimic the native structure with four of our selected perennials. These are species we selected during our inventory phase, species now undergoing germplasm evaluation even as we breed for high seed yield and resistance to seed shatter and pests. Our goal is to build a domestic prairie for grain production. It would be something of a mimic of nature's prairie, though a highly simplified one.

We are not alone on the mimicking front. Dr. Jack Ewel and his colleagues at the University of Florida have mimicked native succession in the tropics. But there are important differences between our work and that of the Ewel team. They have sought to mimic succession (a process) and structure while our efforts are based on phylogeny (relatedness in an evolutionary sense) and structure. The important commonality is mimicry of nature. The Ewel team set out to substitute vine for vine, tree for tree, shrub for shrub. Using in the mimic only plants requiring human intervention, Ewel concludes that nearly always when the structure is imitated the function is granted: high productivity, resistance to pests, and good protection of the soil.

The "mimic approach" employed by The Land Institute and Jack Ewel represents the most extreme experimental category in which nature is used as a standard. Other workers, such as Dr. Steve Gliessman at the University of California at Santa Cruz, employ ecological principles but with little intent to try to so precisely imitate nature's structure. Though some of the work emphasizes diversity over time (crop rotation), it is not necessarily succession. Nevertheless, by featuring diversity, by maintaining ground cover, and by relying on internal sources of nutrients, better control of weeds, diseases, and insects is possible. Nearly all of the good examples of traditional agriculture have employed what we now recognize as sound ecological principles.

I now want to draw on two examples of men from my own field of genetics, two "marginal" scientists with the right idea. The first is Dr. A.L. Hagedoorn, author of a marvelous book entitled *Animal Breeding*, first published in 1939. No one uses this book anymore, but the quality of his insight is unlike that of any other animal geneticist I have known, at least in recent times. I want to use Hagedoorn as an example of the kind of thinking that I hope most agricultural scientists of the future will take more seriously. He had a great distrust of the middlemen in science. He favored direct cooperation between geneticists "with an experimental background, helping to build up the science, and real animal breeders."

In his introduction, Hagedoorn tells us that some of his colleagues encouraged him to split the book into two or three parts so that exhibition stock and utility farm animals would be separated. He refused. He admits the device would have allowed him to go more deeply into some details, but, he says, "I have seen too much stagnation due to specialization."

In addressing the relationship between utility and beauty,

some of Hagedoorn's critics objected to his breeding methods by saying that he would prefer an ugly animal to a beautiful one. He answers these critics by saying that "they have probably never reflected upon the origin of standards of beauty in domestic animals." He continues:

Now, if in selecting swine we restricted our selection absolutely to finding those sires whose offspring graded first-class as butchered hogs, and made profitable gains for the food consumed, the result would be sure to be a breed of swine of great economic value, adapted to local methods of pasturing and feeding, and absolutely right from a butcher's standpoint. Those hogs would have legs, and those legs would be the type of leg a profitable hog should have. Instead of being obliged to *theorize* about the superiority of one sort of leg over the other, we would *know*; and we would think the legs of those ideal pigs were beautiful. Beauty, everywhere, (on an animal) is negative, the beautiful individual being the one that shows no blemishes of any kind, no faults. . . . A beautiful animal is one that shows to perfection the average norm of the best, most useful animals of her breed.

His philosophy is to make each breed of animals as good as possible to fill its niche in the symbiosis-group in which it exists. Hear this:

The animals and plants, and the human race breeding them, belong together in one symbiosis-group, adapted to the country and climatic conditions. It would not be possible to exchange members of one group for those of another group without disturbing the equilibrium.

The Hopis and Zunis of Arizona in the desert have one domestic animal and they have a special kind of maize, and both are specially adapted to life in the desert. The maize can be planted at a great depth in hills containing a number of plants, with considerable distances between hills. The sheep are very small and hardy, adapted to desert conditions by their extremely small size. One sheep weighing a hundred pounds has only one head and one set of legs; two sheep, weighing a hundred pounds together, have two heads and two sets of legs, so that they can be in two different places to hunt the scanty herbs, and for this reason in conditions where the sheep of fifty pounds can just live, a hundred-pound sheep must necessarily starve.

It would not help the Hopis if "good" breeding-stock of bigger size were sent to Arizona and New Mexico to "improve" the local Indian sheep. Those three organisms, the sheep, the drought-resistant maize and the Indian agriculturists belong together in their symbiosis-group in the desert.

Both in Holland and in Germany practically all the goats are kept by very small farmers and labourers. The goats are small and they give very little milk, but they have the advantage of requiring only the coarsest feed—potato peelings, coarse hay cut along the roads, etc. The young females are bred the first autumn. Both in Holland and in some parts of Germany, Swiss goats were imported. Those goats give six and more quarts of milk daily, but they require the best of care, excellent hay and grain, and very good pasture in summer. Goats of this kind were used extensively to "ameliorate" the native labourers' goats. Very soon it was noticed that many people could not afford to keep those goats, especially as the females had to be kept for almost two years before they produced any milk; and as the goats were kept by the very poorest people, who could not afford to spend any money on feed for the goat, the result was a complete failure of the experiment. In one set of two villages in Germany the grading to Swiss goats was tried in village "D," whereas in village "R" no attempt at "improvement" was tried. The result was that, whereas in village "R" all the small labourers kept goats, in village "D" only the *innkeeper*, the *preacher* and the *teacher* kept goats, and the labourers and farmers had given them up as impossible.

When judged at a goat show the large, well-shaped Swiss animals appeared to be vastly superior to the small labourers' goats, but in reality the small goats were vastly superior for the special purpose for which they were kept.

And then he goes on to a discussion about chickens.

> At first sight a breed of poultry, in which the hens lay 200 eggs yearly, is vastly superior to a breed in which the hens lay 50, and yet the cheapest eggs are probably produced by such hens as the native poultry in China and Java, where everybody keeps only a few hens, and nobody ever thinks of giving them any food at all!

My second exemplary breeder, this time a plant breeder, is the late Albert Ahrens from northeastern Nebraska, who, for fifty years, had a private seed corn business, selling to farmers within a fifty-mile radius of his home. He, too, believed that utility and beauty went together. An ear of corn had to "look right," he told me. He looked at an ear of corn with the eye of an art critic. Once he learned that the customers who rotated their crops were not plagued by root worm, he refused to narrow his germplasm to build in resistance to that pest. Instead, he told his customers, "If you want my seed, you'll have to rotate your crops." They did, and both his business and their corn crops thrived until he died a few years ago in his eighties.

Albert Ahrens spent only one year at the University of Nebraska but stayed in touch with the professors the rest of his life. He attributed his well-rounded and independent mind to this university connection and to having "attended high school under strict nuns who were demanding in their teaching of English and botany."

Imagine our continent with thousands of plant and animal breeders practicing their art (and their science) to meet the *regional* necessities of farmers across a less homogenized agricultural landscape. Imagine these men and women breeding crops and livestock for their neighbors, "developing elegant solutions predicated on the uniqueness of place." The range of agricultural possibilities within each ecological island of the mosaic would make maintenance of genetic diversity inevitable. Our high-tech storage facilities, such as the plant germplasm facility at Fort Collins, Colorado, could be mothballed or used for other purposes. That facility, after all, came into existence precisely because industrialized agriculture had little interest in the local adaptations that to some degree were unconsciously maintained by the myriad of farmers across our continent.

The examples just given illustrate what a look to nature can mean for agriculture. This is community ecology in which human-directed arrangements are featured. Which approach will prove the more fruitful toward achieving our goal of a sustainable agriculture—the approach that sets out to mimic, as closely as practical, a natural ecosystem or the "ecological-principles" approach—will probably depend on which landscape or farm is under consideration. Too many variables exist across our ecological mosaic to have a pat answer. We have to keep reminding ourselves of our goals, though—no soil erosion, no chemical contamination of the countryside, no fossil fuel dependency, a return of people to the land, to the small towns and rural communities. In other words, we need to preserve our options. Rich valley land might even be able to experience annual monocultures grown in appropriate rotation with a perennial legume such as alfalfa (diversity over time), while a sloping hillside with shallow soils might require a closer mimicking of the original vegetative structure. My own bias is that, when we don't know, we should employ the mimic approach because by following it out we are more likely to employ undiscovered ecological principles, to take advantage of the natural integrities inherent within the system. Another way to think about it is that by becoming agriculturists who have an eye to natural ecosystems we will increasingly discover answers to questions we have not yet learned to ask.

Ecological principles are abstractions we develop from the phenomena we observe; clearly, both phenomena and abstractions depend on the human as a necessary participant. But our subject now is human community and to what extent human communities can be based on the way a relatively undisturbed natual ecosystem works. As we search for a less extractive and polluting economic order, so that we may fit agriculture into the economy of a sustainable culture, community becomes the locus and metaphor for both agriculture and culture.

When we think of ecology informing agriculture or ecology informing economics, in effect we are expressing an interest in the laws or rules ecologists have elucidated on how ecosystems have worked over the eons. On the one hand, ecologists are accountants who measure what passes across, or through, the boundary of an ecosystem. On the other hand, they also study the dynamics within the boundary, for how materials cycle and how energy flows determine whether, and when, there is a net loss or gain. To do this accounting requires that the accountant be first and foremost a student of boundaries in space and time in order to judge where to place the boundary. Sustainability is a spatial-temporal concept.

The work of the ecologist, accounting at the boundary and measurement of the dynamics within, amounts to a study of ecological community behavior. It can be useful to think of nature's economy as consisting of different species with various degress of complementarity, and to see how that compares to a human economy dominated by different persons with different aptitudes. At this point, however, we have to shift into low gear and proceed with caution, for the analogy has already begun to break down very fast. The first thing to keep in mind is that the essential species in one of nature's land-based ecosystems are the photosynthesizers, plants. Since humans don't photosynthesize but only concentrate, we are forced to think of nature's other concentrators. At this point, we have moved up at least one trophic level, and we can now consider how these economies, the lives of these concentrators, parallel ours. If they are herbivores only, how far do they forage and how is that related to how big they are and at what metabolic levels they operate? Carnivores and omnivores may be two or more steps up the food chain.

Very quickly we see that several biological subdisciplines have become involved here. Unfortunately, few have so far been applied, or in those cases where the relevance of a subdiscipline has been noted, the subdiscipline has not been extensively applied to human economies. Many of the economic differences are so important they may cause us to abandon certain of the ecological principles involved. But in doing this modification, we do *not* abandon the effort to discover commonalities. That is as far as I can go at the moment, but there is another tack that we can and should take.

If we are to look at nature to inform us about sustainable structures and functions in human communities, we must have the audacity to turn back to the paleolithic period and earlier in order to help define what the human is (still) as a social creature. Before agriculture, no more than four hundred generations ago (eight to ten thousand years), essentially all of us were tribal. We may be confident that many, though certainly not all, of the social arrangements in those gathering-hunting bands of twenty-five to fifty and a hundred were survivors of rather intense natural selection. It is also easy to imagine that an environment that shapes us to do certain things for survival purposes made it possible for us to do other things as well. By analogy, a computer is "designed," let's say, to do A, B, C, D, and so forth, and because of its capacity to do these things, it can also do other things, often with ease. The same with humans. We did not evolve riding bicycles, but because of such things as stereoscopic vision and balance—various aspects of our physiology and anatomy that were selected in our pre-agricultural past—today most of us can learn to ride a bike. We might consider bike riding a first-order derivative.

The ability to do some things requires more stretch than the ability to do other things. The closer we stick to our origins in dealing with basic needs (think now in terms of social arrangements), the more likely we are to be successful. We did not evolve in nation states. If we had, bureaucracy would not exist. What we bring from the paleolithic in order to organize the nation state is scant compared to what we need to make it run smoothly. It is a different story when it comes to community. As I see it, community is civilization's upscaling of the gathering-hunting tribe. Why community works has at least as much to do with the way nature shaped us as with the way agriculture and culture have shaped us. Why community works is often a mystery. People often comment that community is fragile, that people want to get away from it. Well, some do, some don't. I am impressed at how people hang on to their places and community life in spite of all the forces, both economic and social, that would destroy community. Public policy with the power of the nation or state behind it is nearly always implemented at the expense of community. Public policy is an abstraction arising out of large social organizations with precedents no more than one hundred centuries old. Community, on the other hand, is a particular, a direct product of our biology, consisting of countless elements, the roots of which reside in social organizations going back to the early humanoids and before. What cultural niceties we have picked up along the way have likely been hard won, too, probably the consequence of disease and death, of ordinary adversity and scarcity.

I am writing in what is left of Matfield Green, a Kansas town of some fifty people situated in a county of a few over 3,000 in an area with thirty-three inches of annual precipitation. It is typical of countless towns throughout the Midwest and Great Plains. People have left, people are leaving, buildings are falling down or burning down. Fourteen of the houses here that do still have people have only one person, usually a widow or widower. I purchased seven rundown houses in town for less than $4,000. Four friends and I purchased, for $5,000, the beautiful brick elementary school, which was built in 1938. It has 10,000 square feet, including a stage and gymnasium. The Land Institute has purchased the high school gymnasium ($4,000) and twelve acres south of town ($6,000). A friend and I have purchased thirty-eight acres north of town from the Santa Fe Railroad for $330 an acre. On this property are a bunkhouse, some large corrals to handle cattle, and around twenty-five acres of never-plowed tallgrass prairie. South of town is an abandoned natural gas booster station with its numerous buildings and facilities. Situated on eighty acres, it had been part of the first long-distance pipelines that delivered natural gas from the large Hugoton field near Amarillo to Kansas City. A neighbor in his late seventies told me that his father had helped dig the basement, using a team of horses and a scoop, around 1929 (contemporary sunlight used to leverage extraction of anciently stored energy). Owned by a major pipeline company, the booster station stands as a silent monument to the extractive economy and a foreshadow of what is to come. A small area has been contaminated with PCBs. Think of the possible practicality and symbolism if this facility, formerly devoted to transporting the high energy demanded by an extractive economy, while at the same time dumping a major environmental pollutant, could be taken over by a community and converted to a facility that would sponsor such renewable technologies as photovoltaic panels or wind machines.

Imagine this human community as an ecosystem, as a locus or primary object of study. We know that much public policy, allegedly implemented in the public interest, is partly responsible for the demise of this community. The question then becomes, how can this human community, like a natural ecosystem community, be protected from that abstraction called "the public." How can both kinds of communities be built back and also protected?

The effort begins, I think, with the sort of inventory and accounting that ecologists have done for natural ecosystems, a kind of accounting seldom if ever done for human-dominated ecosystems. How do we start thinking about what is involved in setting up the books for ecological community accounting that will feature humans? I emphasize accounting because our goal is renewability, sustainability.

About 85 percent of the county in which this small town is situated is never-plowed native prairie. Over the millennia it has featured recycling and has run on sunlight just as a forest or a marsh does. Though the prairie is fenced and is now called pasture and grass, aside from the wilderness areas of our country it is about as close to an original relationship as we will find in any human-managed system. But though the people left in town seem to have a profound affection for the place and for one another, they are as susceptible to the world as anyone else to shopping mall living, to secular materialism in general.

Nevertheless, I have imagined this as a place that could grow bison for meat, as a place where photovoltaic panels could be assembled at the old booster station, where the school could become a gathering place that would be a partial answer to the mall, a place that might attract a few retired people, including professor types, who could bring their pensions, their libraries, and their social security checks to help set up the books for support themselves and to ecological community accounting. Essentially all academic disciplines would be an asset in such an effort, but only if confronted with a broad spectrum of ecological necessities in the face of small-town reality. Much of what must be done will be in conflict with human desires, if not human needs.

But let us allow our imaginations to wander. What *if* bison (or even cattle) could be slaughtered in the little towns on a small scale and become the answer to the massive industrialized Iowa Beef Packers plant at Emporia, a plant to which some area residents drive fifty miles and more in order to earn modest wages and risk carpel tunnel syndrome. What if the manufacture of photovoltaic panels could happen before the eyes of the children of the workers? What if the processing of livestock could happen with those children present? What if those children were allowed to exercise their strong urge to help—to work? What, finally, if shopping malls and Little League were to become less interesting than playing "work-up" with all ages?

One of our major tasks as educators is to expand the imagination about our possibilities. When I walk through the abandoned school with its leaky gym roof and see the solid structure of a beautiful building built in 1938, with stage curtains rotting away and paint peeling off the walls, where one blackboard after another has been carried away as though the building were a slate mine, I am forced to think that the demise of this school has resulted in part from a failure of imagination but more from the tyranny of disregard by something we call "the economy."

I am in that town at this moment, typing in a house that in its history of some seventy-five years has been abandoned more times than anyone can recall. First it was home to a family in a now-abandoned oil field a few miles southeast of here. Then it was hauled intact on a wagon into town in the 1930s, set on a native stone foundation which, when I bought it recently, was crumbling and falling away. Three ceilings supported up to three inches of dirt, much of it from back in the Dust Bowl era. The expanding side walls had traveled so far off the floor joist in the middle that the ceiling was suspended. They had to be pulled back onto the floor joist and held permanently with large screws and plates. Three major holes in the roof had allowed serious water damage to flooring and studs. An oppossum had died under the bathtub long enough ago that only its bones were displayed lying at rest in an arc over the equivalent of half a bale of hay. Two of my seven houses, one empty for nearly twenty years and the other for fifteen, had been walked away from with their refrigerators full. Who knows when the electric meters were pulled—probably a month or two after the houses were vacated and the bill hadn't been paid. To view the contents—eggs and ham, Miracle Whip and pickles, mustard and catsup, milk, cheese, and whatnot—was more like viewing an archaeological find than a repulsive mess.

Where I now sit at my typewriter in that oil field house that relatives and friends have helped remodel, I can see an abandoned lumberyard across the street next to the abandoned hardware store. Out another window but from the same chair is the back of the old creamery that now stores junk (on its way to becoming antiques?). With Clara Jo's retirement, the half-time post office across the street is threatened with closure. From a different window, but the same seat, I can see the bank, which closed in 1929 and paid off ten cents on the dollar. (My nephew recently bought it for $500.) I can see the bank now only because I can look across the vacant lot where the cafe stood before the natural gas booster station shut down. If my front door were open, given the angle at which I sit, I would be looking right at the Hitchin' Post, the only business left, a bar that accommodates local residents and the cowhands who pull up with their pickups and horse trailers with some of the finest saddle mounts anywhere. These horses will quiver and stomp while waiting patiently as their riders stop for lunch or after work to indulge in beer, nuts, and microwave sandwiches and to shoot pool. (The Hitchin' Post lacks running water, so no dishes can be washed. The outhouse is out back.) Around the corner is an abandoned service station. There were once four! Across the street is the former barber shop.

I know that this town and the surrounding farms and ranches did not sponsor perfect people. I keep finding whiskey bottles in old outhouses and garages, stashed between inner and outer walls here and there, these people's local version of a drug culture. I hear the familiar stories of infidelities fifty years back—the overalls on the clothesline hung upside down, the flower pot one day on the left side of the step, the next day on the right—signals to lovers even at the height of the Great Depression when austerity should have tightened the family unit and maybe did. There were the shootings, the failures of justice, the story of the father of a young married woman who shot and killed his son-in-law for hitting his daughter and how charges were never filed. Third-generation bitterness is common. The human drama goes on. This place still doesn't have a lifestyle, never had a lifestyle, but rather livelihoods with ordinary human foibles. Nevertheless, the graveyard now contains the remains of both the cuckolder and the cuckoldee, the shooter and the shot, the drunk and the sober.

This story can be repeated thousands of times across our land, and no telling will deviate even ten percent. I am not sure what should be done here by way of community development. I do believe, however, that community development can begin with putting roofs on buildings that are now leaking, and scabbing in two-by-fours at the bottom of studs that have rotted out around their anchoring nails. I doubt, on the other hand, that a sustainable society can start with a program sent down from Washington or from a Rural Sociology Department in some land grant university, or for that matter as a celebration of Columbus Day.

Locals and most rural sociologists alike believe the answer lies in jobs, *any* jobs so long as they don't pollute too much. Though I am in sympathy with every urgent impulse to meet human welfare, rural America—America!—does *not* need jobs that depend on the extractive economy. We need a way to arrest consumerism. We need a different form of accounting so that both sufficiency and efficiency have *standing* in our minds.

The poets and scientists who counseled that we consult nature would have understood, I think, that we might begin by looking at that old prairie, by remembering who we are as mammals, as primates, as humanoids, as animals struggling to become human by controlling the destructive and unlovely side of our animal nature while we set out to change parts of our still unlovely human mind. The mindscape of the future must have some memory of the ecological arrangements that shaped us and of the social structures that served us well. So many surprises await us even in the next quarter century. My worry is that our context then will be so remote from the ways we survived through the ages that our organizing paradigm will become chaos theory rather than ecology.

4 *Nature as Measure* 🐝

The argument runs like this: We have the poor and starving and we have wilderness. We can't save both. The wilderness advocate: "The poor will be with us always." Even Jesus said it. And besides, their numbers keep multiplying; we can't feed them indefinitely into their next few doublings. "We should do the best we can to feed them, but not at the expense of wilderness, for once wilderness is lost, it's lost forever." And: "The ecosphere gave rise to us. We did not give rise to it. We must keep its creative powers intact." And so on. We might call this the argument of the ecologist. Then there is the argument of the anthropocentrist: "Like it or not, we humans rule the earth." "No wall is high enough or strong enough to keep out the hungry." "Our concepts of justice and righteousness may be very late arrivals on planet earth, but too many of us hold those concepts dear for them to be easily put down." And so on, and so on.

Is there any hope for common ground?

Lynn White in 1967 proposed Saint Francis of Assisi as the patron saint of ecologists. Francis held the radical position that all of creation is holy. Yet nothing in the record shows that he arrived at that position because he was initially endowed with the wilderness psyche of a Henry David Thoreau or a John Muir or an Aldo Leopold. In fact, his entry was from a point at nearly the opposite end of the spectrum—this son of a well-to-do father *chose* poverty. Apparently Francis took seriously the words of Jesus of Nazareth, "If you have done it unto the least of these my brethren, you have done it unto Me." Francis's intimate identification with the "least," his joining the "least," must have prepared him psychologically to be sensitively tuned to all the creation, both the living and the nonliving world. He was a sort of Christian pantheist who believed that birds, flowers, trees, rocks, fire—everything had spiritual standing.

The founder of the most heretical brand of Christianity ever, Francis began his journey with marriage to poverty. That poverty, voluntarily chosen, apparently was the necessary prerequisite for becoming what White calls "the greatest spiritual revolutionary in Western history."[1] His marriage to poverty and his sparing use of the earth's resources led to deep ecological insight. You may remember the legend of the famous Wolf of Gubbio, which had been eating livestock and people. Francis approached the wolf and asked him, in the name of Jesus Christ, to behave himself. The wolf gave signs that he understood. Francis then launched into a description of all the wolf had done, including all the livestock and people he had killed, and pronounced: "You, Brother Wolf, are a thief and a murderer" and therefore "fit for the gallows."[2] The wolf made more signs of understanding, and Francis continued, stating in effect: "I see you have a contrite heart about this matter, and if you promise to behave, I'll see to it that you are fed." The wolf showed signs that he promised, and the legend has it that the wolf came to town, went in and out of people's houses as a kind of pet, and lived that way for two years before dying. The townspeople all fed the wolf, scratched its ears, and so on. Its last days were spent enjoying the good life at the dog food bowls of Gubbio.

In light of this story, I once thought that Saint Francis might more properly be regarded as the patron saint of domesticators. Anyone able to encourage a wolf to stop behaving as we believe a wolf must behave, given everything from its enzyme system to its fangs, is not likely to be regarded as an ecologist, let alone a patron saint for such.

About three o'clock one morning a few winters ago, the barking of our two dogs woke me up. Soon there was hissing and growling and more barking right on the back porch. I went outside and saw a raccoon cowering under a step stool by the dog food bowl. I chased him away with a stick, sicced our border collie, Molly, on him and went back to bed. I confidently went to sleep. A few minutes later I heard more barking, more hissing, more growling. Once again I went outside, and the dog and I ran the coon off. Back to bed, and sure enough the same story. This time, however, I left the porch light on for my visitor. I lay in bed and listened; there was no more ruckus. I felt pleased with myself for solving the problem with a light switch thanks to my knowledge about nocturnal animals, and I went to sleep. The next morning I headed out the back door to begin the day's work, and there in the box of tinder on a table was the coon, sleeping away. Both dogs were asleep under the table. Each time I returned to the house during the day, I expected to see that the coon had gone. But it didn't leave. And as it slept that day, the dogs would walk by, look up, and sniff, well on toward accepting their new fellow resident.

What was going on here? We are taught to consider recent changes when something unusual breaks a pattern. And so I may have an answer. Late in the afternoons that winter I had been going to my woods with a chain saw, some gasoline, and matches to burn brush. I had cleared out most of the box-elder trees that had grown up there over the last forty years. They were early-stage succession trees that had mostly covered the area where a former tenant had logged out all the walnuts and burr oaks. The box elders are the first to green up in spring, and they accommodate woodpeckers. They are no good for lumber, and though we burn them, they are very low in fuel value. I want to accelerate succession by planting some walnuts and oaks for my grandchildren. Nearly every box elder I took down was hollow. The woods are less than a quarter of a mile from my house along the Smoky Hill River. I suspect that I had destroyed the home of Friar Coon, who, looking for a new home, simply moved into mine.

Back to Gubbio. Why was the thieving, murdering wolf forgiven and granted a life of ease at the dog food bowls? I suspect that the ecological context necessary to accommodate proper wolfhood around Gubbio had been destroyed, that the usual predator-prey relationships had been disrupted, that Francis realized this and that the wolf was hungry. If so, his deeper ecological insight came perhaps from his respect and love for nature. He could, after all, have accommodated the posse of armed and fearful citizens out to eliminate the killer wolf, but instead his love for nature—in turn at least partly derived from his identification with poverty—made him forgiving and compassionate.

But there is still another item on our agenda for discussion. Was Saint Francis engaged in an act of domestication? It appears so; but if harmony with nature is what we seek, should we not be willing to achieve harmony any way we can? The wolf's tame behavior demonstrated that nature is not rigid. Humanity (Francis) reached out to nature (the wolf), and nature responded. If we insist that wild nature be rigid, we deny one of its most important properties—resilience. The unanswered question here is: Once a wolf has come to town, can it ever really return to the wild? Life at a dog bowl in Gubbio may be easier than life that relies on fangs and the occasional berry. If the wolf is unable to return, then that particular wolf is a fallen creature. Would it matter that the wolf was made a fallen creature initially by a fallen ecological context man had brought about? Grizzlies in the garbage at Yellowstone and elephants in African dump heaps come to mind as modern victims of the same problem. A wolf or any other creature unable to return is a fallen animal dependent on fallen humanity. And so another ethical question comes into our agenda:

What right do we have to create a fallen world for other species when we know that life at the dog food bowl is a second-best life? And following an analogous thought, we find that our crops and livestock represent fallen species, fallen to accommodate fallen humanity.

The stories about the wolf in Gubbio and the coon on the porch illustrate the problem of trying to use nature as measure, for in these examples the interpenetration of the domestic and the wild is total. During the last five hundred years or so, the ratio of the domestic to the wild has been so altered, especially in the Western Hemisphere, that wilderness itself has become an artifact of civilization. Only civilization can save wilderness now; the wild that produced us, that we were dependent upon is now our dependent. We pay homage to wildness in the United States by regarding pristine wilderness as a kind of saint. That creates more problems.

From time to time I have heard of Christians who go into the quiet and set-aside peacefulness of a church and move to the image of an appropriate saint to stand or kneel, light a candle, whisper some words, and return to the noisy world of fellow beings with a sense of peace. Maybe they've touched a useful and virtuous base, and maybe only just in time. Isolated, preserved virtue has been sought out and appealed to for succor and support, though for exactly what only the supplicant knows.

In an analogous way, countless advocates of wilderness go to their set-aside wild places of quiet and peace and find spiritual renewal, then return to the noisy world to lobby and lecture, to send money to maintain those wild shrines. And in doing so, many of them no doubt feel quite pious and, as it were, "saved." I do not want to put down either the utility of saints or the sincerity of advocates of wilderness. But as even good Christians generally go astray on a regular basis, so advocates of wilderness too often have very little to say in protest of, for example, the spread of lethal farm chemicals over more than half a million square miles of the best agricultural land in the world, or soil erosion, or . . . . Well, to treat wilderness as a holy shrine and Kansas or East Saint Louis as terrain of an altogether different sort is a form of schizophrenia. Either all the earth is holy or none is. Either every square foot of it deserves our respect or none does.

Can Earth First! activists or Deep Ecologists be as interested in cleaning up East Saint Louis as they are in defending wilderness? Can Earth First! activists be as fervent in defending a farmer's soil conservation effort or chemical-free crop rotation as they are in spiking a tree or putting sugar in the fuel tank of a bulldozer?

It is possible to love a small acreage in Kansas as much as John Muir loved the entire Sierra Nevada. This is fortunate, for the wilderness of the Sierra will disappear unless little pieces of nonwilderness become intensely loved by lots of people. In other words, Harlem and East Saint Louis and Iowa and Kansas and the rest of the world where wilderness has been destroyed must come to be loved by enough of us, or wilderness too is doomed. Suddenly we see we are dealing with a range of issues. Saint Francis's *entire* life becomes an important example. People who struggle for social justice by working with the poor in cities and those who labor to prevent soil erosion and save the family farm are suddenly on the same side as the wilderness advocate. All have joined in the same fight.

Wendell Berry once said to me, "It will be an awful thing if we quit being human before we become extinct." When compassion, a human characteristic, is denied in the interest of salvation of the earth, we deny the full dimension of the humanity we must have to save the earth. Time is short. We have too much to do. But concern for all carries with it the solution we need for the preservation of the One—the Ecosphere.

I mentioned earlier that Francis was a sort of pantheist. It did not matter whether it was an object, a creature, or a phenomenon in nature; all had a selfhood, an otherness which he honored. He had his own sort of conversations with nature, and they must have numbered in the thousands during his lifetime. Once he was near blind and in need of a cauterization from the ear to the eyebrow. He spoke to the fire as his brother and told the fire, "The Most High has created you strong, beautiful, and useful giving you a dazzling presence, which all other creatures envy. Be kind and courteous to me in this trance. I beg the Lord to cause you to temper your strength, so that by burning me gently I may tolerate you." Legend has it that Brother Fire did have mercy on Francis.

To most of us it seems ridiculous that all of nature should have such a standing as it did with Francis. There must have been a time, certainly when pantheism was widespread, before Christianity moved into Europe, that something close to his attitude was common. It is a far cry from Francis's experience to the late twentieth century, when nature has become commodity—human property to be bought and sold in the market. We European conquerors and settlers were amused at the Native Americans who could not quite comprehend how land could be sold. Now we can empathize with them, for many of us have a difficult time comprehending the serious proposal to sell much of what is left of the global commons, such as air. A serious proposal to sell air had its beginnings around 1960 in an article entitled "The Problem of Social Cost" by Dr. Ronald Coase.[3] Some three decades later, this man received a Nobel prize for such "forward thinking." Now serious economic scholars are examining his idea, especially those who have seen the failings of government when expensive pollution-control technologies are mandated or when industries responsible for emissions are penalized.

Professor Coase states that pollution debates in reality amount to conflicts over the use of scarce resources. Factories want to dump soot into the air, thereby externalizing the cost; but residents want clean air to breath. He proposes that we solve this conflict in the market, where other conflicts are resolved. How to do this? Assign ownership rights and let people trade.

Jerome Ellig, a professor at George Mason University, has studied Coase's ideas and says of them:

Suppose the surrounding homeowners owned the air, so that the factory had to pay them if it wanted to pollute. For a price, the neighbors might tolerate some soot, but at some point, it would be cheaper for the factory to install pollution controls than to buy the right to pollute. If clean air is sufficiently valuable to the factory's neighbors, the factory will have to install the pollution controls, because it is cheaper to control the pollution than to pay compensation to the neighbors.

Now suppose the factory owned the right to pollute the air, and the surrounding homeowners had to pay the factory to clean up its act. If the value they place on clean air exceeds the cost of the pollution control equipment, they will be willing to pay the factory to install the equipment.[4]

The assumption is that pollution control equipment *will* be installed because clearly defined property rights encourage both the factory owners and the neighbors to strike a deal. I don't know whether it is Ellig reporting about Coase or Coase himself who says that "The only reason such an agreement will not occur is if transaction costs prevent the factory owner and the community from sitting down and negotiating. For instance, in a large neighborhood, the cost of organizing the residents and finding out how highly they value clean air may be very large."

But there are other obstacles. For starters, what about the factory moving to a poorer neighborhood or a poorer part of the country or a poorer part of the world? We've already seen that happen, though Ellig and Coase would say in all likelihood that the moves came about because governments were the regulators. They believe that the role of government should be to "make sure that property rights to use land, air and water are clearly defined, then eliminate or reduce the transaction costs that prevent polluters and communities from striking deals. In this way, the decisions of the citizens most affected by pollution, rather then distant government officials, determine how much, if any, pollution will be allowed."

So now we have light! There is the air, a common good of nature; appropriate it, assign ownership to it, move it into the market economy. The assumption, of course, is that we all have total knowledge, that we will act on the basis of our knowledge to meet not only our present needs but also those of generations to come. But the real knowledge of everyone who has thought about the problem of pollution or has had practical experience wrestling with it is that generally the factory can control the data better than the receivers of the pollutants can uncover the data.

Furthermore, what is not addressed is how to determine the boundary of the pollutant. Take, for example, chlorofluorocarbons. They move into the atmosphere and, on average, experience approximately fifty thousand ozone-destroying reactions before breaking down. As mentioned earlier, a scant quarter of a century ago, CFCs would have been on nearly every chemist's shining short list. Not only was the result of our experiment with CFCs unforeseen, it was probably unforeseeable.

Ronald Coase's assumption has to be based on the idea that when we run an experiment our knowledge is adequate—that is, that if there is any remote possibility of harm to the environment, our knowledge will be sufficiently complete that we can make every necessary correction.

Before economists or anyone else becomes too excited

about the recommendations of Nobel Laureate Coase, how about just one master's or doctor's thesis simply compiling a list of the major factories that have gone out of existence in the last five years? I say major factories because of the sort of financial commitment that must have gone into them and therefore the amount of information that must have been brought to bear before the financial commitment was made. Or how about an analysis of why we lack the ability to make a sound decision on matters surrounding nuclear power?

When the values and methodology of the bookkeepers of business generate the most creative environmental paradigms, it is time to turn to nature's economy for more and better ideas.

Before I launch into that discussion, however, an important qualifier is necessary. When the native peoples of this continent were its only human inhabitants, they managed, to some degree, almost all of it. Much of the eastern deciduous forest experienced twice-yearly burning of the undergrowth. Sometimes these burnings were quite extensive. Verrazzano in 1524 described Rhode Island as a place of meadows that extended from seventy-five to ninety miles with no trees. Out on the prairies, nature through lightning and Indians with torches set fires. The grass would green up faster in the spring where the old fuel had been burned. On the eastern edges of the prairies, fire kept woody vegetation back. The grasslands had likely evolved to invite naturally occurring fire. But the natives used fire as a technique to drive out game, perhaps even before the horse arrived. So this continent has experienced extensive land management probably almost as long as humans have lived here. To manage does not imply that we cannot become native.

In our own work at The Land Institute, ecology is our primary field of interest because nature is our standard, the model we use as we design our experiments. Nature as standard, as "measure," is not a new idea. As Wendell Berry points out, the notion goes back to at least two thousand years before Jesus of Nazareth. In a memorable speech delivered at the dedication of our new greenhouse at The Land Institute in 1988, Wendell Berry traced the literary and scientific history of our work.[5] He began by quoting Job: "Ask now the beasts and they shall teach thee, and the fowls of the air, and they shall tell thee: Or speak to the earth, and it shall teach thee; and the fishes of the sea shall declare unto thee." Then Virgil who, at the beginning of *The Georgics* (36-29 B.C.), instructs us that "before we plow an unfamiliar patch / It is well to be informed about the winds, / About the variations in the sky, / The native traits and habits of the place, / What each locale permits, and what denies."[6] Toward the end of the sixteenth century, Edmund Spenser called nature "the equall mother" of all creatures, who "knittest each to each, as brother unto brother." Spenser also saw nature as the instructor of creatures and the ultimate earthly judge of their behavior. Shakespeare, in *As You Like It*, has the forest in the role of teacher and judge. Milton, in *Comus*, has the Lady say of nature, "she, good cateress, / Means her provision only to the good / That live according to her sober laws / And holy dictate of spare Temperence." Finally, Alexander Pope, in his *Epistle to Burlington*, counseled gardeners to "let Nature never be forgot" and to "Consult the Genius of the Place in all."

After Pope, Berry points out, this theme of a practical harmony between man and nature departs from English poetry. Later poets see nature and humanity radically divided. A practical harmony between land and people was not on their agenda. The romantic poets made so central the human mind that nature became less a reality to be dealt with in a practical way and more what Wendell Berry refers to as a "reservoir of symbols."[7]

We have largely ignored this literary tradition, of course. Nevertheless I cannot help but wonder what the consequences would have been if the settlers and children of settlers whose plowing of the Great Plains in the 'teens and twenties gave us the Dust Bowl of the thirties had heeded Virgil's admonition that "before we plow an unfamiliar patch it is well to be informed about the winds." What of Milton's insight about the good cateress who "means her provision only to the good / That live according to her sober laws / And holy dictate of spare Temperence"? Virgil was writing about prudent agricultural practices, Milton about prudent consumption, the spare use of nature's fruits. For both, nature gave the measure, the standard, the lesson.

Our nation has not yet even begun seriously building a *science* of agricultural sustainability with nature as the measure. A few scientists have spoken in terms that echo the poets. In the paper already quoted, Wendell Berry noted that after the theme "nature as the measure" went underground among the poets in the last century, it next surfaced among the agricultural writers who had a scientific bent. Liberty Hyde Bailey's *The Outlook to Nature* appeared in 1905.[8] That grand old Cornell dean described nature as "the norm": "If nature is the norm then the necessity for correcting and amending abuses of civilization becomes baldly apparent by very contrast." He continues: "The return to nature affords the very means of acquiring the incentive and energy for ambitious and constructive work of a high order." Later, in *The Holy Earth* (1915), Bailey advanced the notion that "a good part of agriculture is to learn how to adapt one's work to nature. . . . To live in right relation with his natural conditions is one of the first lessons that a wise farmer or any other wise man learns."[9]

Sir Albert Howard published *An Agricultural Testament* in 1940. Howard thought we should farm as the forest does for nature constitutes the "supreme farmers": "The main characteristic of Nature's farming can therefore be summed up in a few words. Mother earth never attempts to farm without live stock; she always raises mixed crops; great pains are taken to preserve the soil and to prevent erosion; the mixed vegetable and animal wastes are converted into humus; there is no waste; the processes of growth and the processes of decay balance one another; ample provision is made to maintain large reserves of fertility; the greatest care is taken to store the rainfall; both plants and animals are left to protect themselves against disease."[10]

Earlier, in 1929, J. Russell Smith in his *Tree Crops* stated that "farming should fit the land." He was disturbed by the destruction of the hills because "man has carried to the hills the agriculture of the flat plain."[11] (An agriculture modeled on the prairie featuring perennials would make possible grain harvest on hillsides.)

Is our current emphasis on sustainable agriculture at The Land Institute part of a succession in which nature is the measure? It is, in a way, for as Wendell Berry said about the poets and scientists he quoted, there is a succession in thought but only in the familial and communal handing down in the agrarian common culture, not in the formal culture, where it exists only as a series. It is interesting but not surprising that the common culture had a succession but teachers and students in the literary or scientific tradition could only manage to provide a series. Why they never built on the writings of those who had gone before is an important question, one that needs to be answered.

But there is more to the problem. Those who popped up from that common culture to form that series, whether poets or scientists, did not make us *their* successors, or, put another way, we have not made ourselves their successors. So here is the challenge. We have a chance to begin to build that formal succession now. For now, by trying to understand agriculture in its own terms, we see what has happened and we can build on the science of ecology and evolutionary biology. But because our work gets down to experiments and data, we risk falling into Baconian-Cartesian reductionism. We need more people who will show us the *practical* possibility of a research agenda based on a marriage of agriculture and ecology. That agenda will require a push from those who, after examining the assumptions of modern agriculture versus what nature has to offer, decide in favor of learning from nature's wisdom.

We look to natural ecosystems because they have featured recycling of essentially all materials and have run on sunlight. I say "feature" because they have not been perfect in those recycling efforts. For that matter, not all life forms are powered by the sun. The exceptions, however, are trivial. Ecological standards based on studies of ecosystems that have experienced minimum human impact provide us with our best understanding of how the world worked during the hundreds of millions of years before humans arrived.

With this in mind, I have two stories. The first amounts to an ecological comparison of two land tracts. In 1933, a graduate student at the University of Nebraska carried out a research project near Lincoln in which he compared an upland, never-plowed prairie with an adjacent field of winter wheat. Prairie and wheat were growing on the same soil type, but when moisture fell, 8.7 percent ran off the wheat field while only 1.2 percent ran off the prairie. It turned out to be the driest year on record. All the wheat plants died, while the deep-penetrating perennial roots of the prairie survived. The upshot of this story is that prairie is "designed" to receive water efficiently and then to allocate that water carefully. An average day in the spring would find the wheat field losing nearly twenty-one tons of water per acre; on the same day the prairie would lose only a little over thirteen tons per acre. This economy was produced by such mechanisms as moderating wind speed and keeping temperature as low as possible. There are other interesting comparisons in that study, but let's stick with water.

For the second story, let's leave Nebraska and go to the tropics, to a tropical rain forest in Costa Rica where Jack Ewel and his colleagues from the University of Florida have compared agricultural fields with the surrounding forest. Here, water can be the nemesis of fertility, for when the forest is destroyed, valuable nutrients are leached downward. A rain forest, on the other hand, is "designed" to pump that water back to the atmosphere with great efficiency.

Thus with respect to water management, we have in these examples two opposite ecosystems. Both are keyed to the needs of their places. Nature's prairie holds water; the wheat field loses it more rapidly. Nature's tropical rain forest gets rid of water; agricultural patches in the tropical rain forest lose fertility because not enough water is intercepted and pumped away.

These stories not only describe realities in nature, they provide lessons with which we humans must come to terms. First of all, the stories illustrate that when we humans mess around with an ecosystem, we tend to invert what nature does well. Just as bad, we tend to ignore the question of why nature features ecological mosaics that, until disturbed for human purposes, provide, in the words of John Todd, "elegant solutions predicated on the uniqueness of [each] place."[12] To much too large a degree this lesson has been ignored as agriculture,

particularly industrial agriculture, tends toward the homogen-
ization of landscapes.

In Howard T. Odum's *Environment, Power and Society* there
appears a memorable figure with three important drawings.[13]
The first shows a rectangular microscope slide containing numerous photosynthesizers, microscopic algal cells (light fixers)
more or less evenly spread across the slide. The source was
Florida pond water. Concentrators in the form of tube animals
are sprinkled here and there across the slide. The second drawing is also a rectangle, a sketch of an aerial view of a rural Kansas
landscape showing the photosynthesizers as crop plants that fix
sunlight and the farmsteads as the concentrators of that solar
energy. I imagine this view of Kansas reflects reality before the
intense industrialization of agriculture. The third drawing is of a
tropical rain forest canopy that carries on photosynthesis, and
here the trunks and limbs are the concentrators.

What is notable about the sketches is that the relationship of photosynthetic area to concentrators is about the same
when reduced to a common scale. This is only a very general
reality, for whales travel vast distances and so do birds. The
bison migrated, and so have and do many indigenous peoples.
But in a world that is filling up with people and their technological demands, if we stick to sun power, one is forced to
think about the limits of the area a concentrator can utilize in
most cases. From here on it takes little imagination to specu-
late on the demise of the family farm and rural communities.
The "get big or get out" era following World War II was likely
the consequence as much of the introduction of fossil fuel sub-

sidy into the agricultural system as of the economics that attended that perception of reality. In other words, farm policy originating in Washington may have been less responsible for the reduction in rural life than the disruption due to the fossil fuel subsidy.

A prevailing notion is that the tractor was more intrinsically desirable than the draft animal. That may be, but it is at least worth remembering how many draft-animal-using farmers hung onto horse and mule farming for as long as they could. Those who loved and could get along with horses could not compete with those enamored by the machine. In energy language, systems powered by contemporary sunlight could not compete with systems powered by anciently stored energy.

When I think of those drawings of H.T. Odum and the possibility of a sunpowered agriculture with diesel tractors burning vegetable oils, I wonder if we will not be forced to return to a relationship between photosynthesizer and concentrator as illustrated in Odum's figure. And what does this mean about the necessity to repopulate much of the countryside and most of the rural communities? When we were migratory people maybe some of us did follow the sun over a large area. With agriculture and an exploding population assuming a sun-powered technological array, it seems that we will be forced into more restricted areas both to gather our food and to sponsor our technology.

For many a scientist, there is *the* memorable field trip, one that sticks out from all the rest. Mine lasted three days in

September of 1985 near Comptche in Mendocino County, California. Hans Jenny and his friend Arnold Schultz, a forestry professor at the University of California, Berkeley, led Saskatchewan ecologist J. Stan Rowe and me up and down what is called the ecological staircase of Mendocino. It was a trip in which any penchant toward eco-fundamentalism was sure to suffer. At least mine did.

To understand an ecological staircase, imagine that you are on the beach looking eastward with your back to the Pacific. If you were several hundred feet tall, and especially if no vegetation was in the way, you would have no trouble seeing five stair treads, each one representing uplift due to the Pacific plate sliding under the continental plate every 100,000 years. Before I started up the Mendocino staircase, I was a firm believer that any natural ecosystem must invariably improve, and by that I mean add topsoil and increase in stability and maybe in diversity; or, if the ecosystem should happen not to improve, at least it would stay good indefinitely. By the time we headed back toward Berkeley in the car, the pillars of my ecological understanding had been shaken.

My concerns grew over the next several weeks. Finally, about four months after the field trip, a letter came from Hans saying that he was not aware of such a concept as steadily improving ecosystems. He said that such a "sunshiney belief rests on a neglect to appreciate the soil as a dynamic—either improving or degrading—vital component of land ecosystems." There was little comfort in the fact that I had had it half right.

In that same letter he expressed his concern as to whether he and Arnold had presented Stan and me with "sufficient physical evidence that the decline in soil and vegetation from the redwood-Douglas fir forest on the second terrace to the pygmy forest is a natural sequence." (The third terrace is a transition step to the pygmy forest on steps five and six.) Plant ecologists had, after all, designated the redwood-fir forest a climatic climax (so long as the climate was constant the vegetative structure would not change) and the pygmy forest an edaphic climax (relating to soil). In Hans's view, ecologists had designated two different worlds, two different ecosystems, "not realizing that the two ecosystems might be on the same time arrow, merely separated by a long time interval."

Fundamentalism of any variety tends to die hard. Staring into a soil pit dug into the fourth terrace, I could sympathize with the churchmen who refused to look through Galileo's telescope. Even there, with the evidence before me, I protested, saying that good farming *can* improve the soil. "Yes," Hans said, but "the extent depends on what kind of a soil, virgin or depleted, the farmer begins with." He thought it would be difficult to improve a good virgin Iowa prairie soil by soil management techniques, except perhaps by applying nitrogen, phosphorus, and potassium.

This was the beginning of an important lesson for me. Ever since then I have burdened myself and my students with the question: Why should a look to nature, as we work out our relationship to the earth, provide us with easy absolutes? Nature may or may not share human interests. It is we who choose to make nature our standard or measure for agriculture instead of trying to understand agriculture on its own terms. It is also we, not nature, who have concepts and notions of good and bad. Few humans, in comparing the luxuriant redwood-fir forest to the pygmy forest, would not think that the latter represents a deterioration or decline. But Hans insisted that "nature might call it a biological improvement, an adaptation of vegetation to a changing substrate." This, by the way, raises another question: Why are there not then pygmy forests or pygmy prairies, or pygmy whatevers, all over the world? The answer is: because we have disturbances—glaciers, volcanoes, mountain formation due to uplift—that seem destructive yet are sources for eventual soil formation. But this requires very long time periods, so soil is as much of a nonrenewable resource as oil.

John Cobb, the Whiteheadian philosopher, has written me that "We cannot learn from [nature] except as we ask questions and we have to be ready to have the questions revised by the answers." This is a near parallel to a statement made by Wendell Berry, who says that we need a conversation with nature. We should favor a highly interactive approach in which, as John Cobb says, "we neither try to impose our categories nor merely adapt to what is."

Perhaps it is the spirit in which we ask our questions that ultimately will determine our fate. Rather than ask what nature requires of us here, we mostly ask what we can get away with. The latter is a childish question. The former does not necessarily assume that nature has any moral authority, but it does not rule it out either, and it implies, I think, that *we* have moral responsibility. We will probably never know enough to know whether nature is moral. Unable to know, we must trust in this source of humility, along with lessons that come from those steps, the oldest reaching back 500,000 years.

The material that arrives daily from outer space is so trivial compared to what is already here that it doesn't count. Fossil fuels are the consequence of relatively minor "failures" of life forms to experience total respiration. Nature doesn't totally recycle everything. Volcanoes add new material to the surface each year, and glaciers grind away at rock with varying degrees of hardness to give us material for soil. Tectonic plates are at work heaving up material more or less all the time, creating acres for real estate; but no real estate office would ever await such a land arise.

Nature's economy in a forest or a prairie is more important for us to look at in thinking of the human enterprise than these long-term replenishments. This greater economy that we seldom acknowledge we share with all creatures, and we have no choice in the matter. Its processes should be of great interest and importance to us.

The factuality of nature cannot be our total morality, but by being ignorant of nature we are ignorant of our limits as well as our possibilities. In the words of Stephen Jay Gould, "Why should a process that regulated 3.5 billion years of living creatures without explicit ethical systems provide all the answers for a species that evolved only a geological second ago, and then change the rules by introducing such new and interesting concepts as justice and righteousness?"

We are not looking to nature for all the answers. The limits as well as the hideousness of social Darwinism of the last century, continuing into our own century, are well known. Pope's phrase is the best: "Consult the genius of the place in all." The key word is "consult." Such a fanatical look to nature, in fact, could justify many of the procedures of modern economics. After all, modern economic theory validates the idea that individuals should act to optimize their own interests. Or in the words of Herman Daly and John Cobb, we are to be self-regarding rather than other-regarding. When we look at a prairie or forest ecosystem, "selfishness" may *appear* to be the case; but when we read of the behavior of many of our primate relatives we see altruism, which may be a hidden form of self-ishness. But that is another discussion.

The prairie or the forest can exist in a dynamic equilibrium, allowing several species to flourish, because there is no other species present with the technological power of the human. If our cue is taken from some aspects of primate behavior, then we are ready for a discussion of the merits of Calvinism and Catholicism. Daly and Cobb point out that Calvinism encourages other-regarding behavior as truly Christian, but they warn against believing too readily in its reality. Catholicism regards other-regarding behavior as a natural virtue of humanity.

The economists have come along and taught us to believe that checks on self-interest are not only unnecessary but

harmful. In their minds, self-interest behavior is rational behavior. Now that this ethos has become the dominant force at
work in the market, wittingly or not, given the technological
array that has popped up, the earth, including countless life
forms, has become a mine and an overflowing sink for our
wastes. That is not the way a healthy prairie works, where
wastes become ecological capital.

5 *Becoming Native to Our Places* 🐾

It seems to be a characteristic of life that no matter what the level of organization, the juvenile stage is characterized by an excess of potential energy and an inefficiency in use of that energy. This seems to be as true of the early stages of an ecosystem as of a teenager. But we have seldom considered a corollary—that an excess of potential energy can *generate* a juvenile condition. The industrial revolution really hit its stride after World War II. It was only then that we became a truly affluent society. The Depression and the war contributed to making us a disciplined people, but after the war economic growth and invention really took off. We came to believe that comfort and security were the solutions to the human condition.

But what this excess of potential energy has yielded us, beyond the throughput of goods, is a decrease in our maturity. Our culture is now like a time machine running backward. National polls frequently show that when the issues are framed as value questions, the public will give what in my view is the responsible answer. Then they'll vote otherwise. We saw this during the Reagan and Bush years. Environment gets a high

rating because it is the right answer, but people want government to do it without raising taxes or having a national discussion about getting rid of the automobile. It is reminiscent of a
child who can give the answer the parents want and then goes
on and does what he or she wants to do.

This is not what Madison and the founding fathers expected. They believed the maturity of people's judgment
would expand. Worried about corruption, they assumed that
eventually our judgment would be larger, more diverse, and
therefore more stable. Instead, we have gone the other way.
We have become a more juvenile culture. We have become a
childish "me, me, me" culture with fifteen-second attention
spans. The global village that television was supposed to bring
is less a village then a playground. We'd rather gossip about
President Clinton's sex life than talk about the issues. And so
few of the issues are really being dealt with. We seem satisfied
to keep tossing around that vague term "environment" without
talking about our relationship with nature. The destruction of
wilderness is not even a secondary consideration. Community
destruction is scarcely mentioned. Destruction of our agricultural communities may as well go unnoticed; little is done
about it. Widespread if not universal child neglect is less discussed than "the economy."

Nearly all of the suggestions for change are off the mark.
We educate kids to take tests. We make the assumption that
better organized education will be better education. But what of
the content? Teachers don't even know how to talk about community responsibility. Little attempt is made to pass on our cultural inheritance, and our moral and religious traditions are ne-

glected except in the shallow "family values" arguments. In our universities there is good reason to believe that the Declaration of Independence would not be passed by university professors if it were brought to a vote today. Unlike those who signed that document, most modern scholars are less servants of the people.

A necessary part of our intelligence is on the line as the oral tradition becomes less and less important. There was a time throughout our land when it was common for stories to be told and retold, a most valuable exercise, for the story retold is the story reexamined over and over again at different levels of intellectual and emotional growth. *Huck Finn* read at the fifth-grade level is different from *Huck Finn* read in high school or college or as a young parent or grandparent. That is true with almost any story. But "news" as displayed on television appears once only, unlike the story in the oral tradition with its many levels of meaning.

Entire neighborhoods are more accessible to the world than their members are to one another. Is this part of our nature? It is always easier to think of a better way to produce food or a consumer item than to think of how to avoid using that food or that gadget wastefully. We waste, I believe, largely because of our fallen condition. We employ human cleverness to make the earth yield an unbounded technological array, which in turn produces countless more technologies, more things. In agriculture, we hot-wire the landscape, bypassing nature's control devices. We do this in the face of abundant evidence that we are destroying our habitat because of our "unwitting accessibility" to the world.[1] I should explain my interpretation of this phrase.

A few years ago on the last page of *Life* magazine I saw a memorable photograph of a near-naked, well muscled tribesman of Indonesian New Guinea who was staring at a parked airplane in a jungle clearing. The caption noted the Indonesian government's attempt to bring such "savages" into the money economy. A stand had been set up at the edge of the jungle and was reportedly doing a brisk business in beer, soda pop, and tennis shoes. We can imagine what must have followed for the members of the tribe, what the wages of their "sin," their "fall," must have been—decaying teeth, anxiety in a money system, destruction of their social structure. If they were like what we know of most so-called primitive peoples, then in spite of its hierarchical structure, their society was much more egalitarian than today's industrialized societies.

Unlike Adam and Eve, the New Guinea tribesmen received no explicit commandment to avoid the goodies of civilization. They simply accepted, unwittingly, the proffered accessibility to beer, soda pop, and tennis shoes. In Genesis, the primal sin involves disobedience, an exercise of free will. In our modern non-Paradise version, "original sin" is our unwitting acceptance of the material things of the world. I perceive this to be the largest threat to our planet and to our ability to accept nature-as-measure.

In *Beyond the Hundredth Meridian* (1953), Wallace Stegner describes the breakdown of American Indian culture:

However sympathetically or even sentimentally a white American viewed the Indian, the industrial culture was certain to eat away at the tribal cultures like lye. One's attitude might vary, but the fact went on regardless. What destroyed the Indian was not primarily political greed, land hunger, or military power, not the white man's germs or the white man's rum. What destroyed him was the manufactured products of a culture, iron and steel, guns, needles, woolen cloth, *things that once possessed could not be done without.* [Italics added.]

It was not the continuity of the Indian race that failed; what failed was the continuity of the diverse tribal cultures. These exist now only in scattered, degenerated reservation fragments among such notably resistant peoples as the Pueblo and Navajo of the final, persistent Indian Country. And here what has protected them is aridity, the difficulties in the way of dense white settlement, the accident of relative isolation, as much as the stability of their own institutions. Even here a Hopi dancer with tortoise shells on his calves and turquoise on his neck and wrists and a kirtle of fine traditional weave around his loins may wear down his back as an amulet a nickel-plated Ingersoll watch, or a Purple Heart medal won in a white man's war. Even here, in Monument Valley where not one Navajo in ten speaks any English, squaws may herd their sheep through the shadscale and rabbitbrush in brown and white saddle shoes and Hollywood sunglasses, or gather under a juniper for gossip and bubblegum. The lye still corrodes even the resistant cultures.

This reality—things once possessed that cannot be done without—is so powerful that it occupies our unconscious. And yet we know that nature, in Milton's words, "means her provision only to the good / That live according to her sober laws / And holy dictate of spare Temperence."

At work on my houses in Matfield Green, I've had great fun tearing off the porches and cleaning up the yards. But it has been sad, as well, going through the abandoned belongings of families who lived out their lives in this beautiful, well-watered, fertile setting. In an upstairs bedroom, I came across a dusty but beautiful blue padded box labeled "Old Programs—New Century Club." Most of the programs from 1923 to 1964 were there. Each listed the officers, the Club Flower (sweet pea), the Club Colors (pink and white), and the Club Motto ("Just Be Glad"). The programs for each year were gathered under one cover and nearly always dedicated to some local woman who was special in some way.

Each month the women were to comment on such subjects as canning, jokes, memory gems, a magazine article, guest poems, flower culture, misused words, birds, and so on. The May 1936 program was a debate: "Resolved that movies are detrimental to the young generation." The August 1936 program was dedicated to coping with the heat. Roll call was "Hot Weather Drinks"; next came "Suggestions for Hot Weather Lunches"; a Mrs. Rogler offered "Ways of Keeping Cool."

The June roll call in 1929 was "The Disease I Fear Most." That was eleven years after the great flu epidemic. Children were still dying in those days of diphtheria, whooping cough, scarlet fever, pneumonia. On August 20, the roll call question was "What do you consider the most essential to good citizenship?" In September that year it was "Birds of our country." The program was on the mourning dove.

What became of it all?

From 1923 to 1930 the program covers were beautiful, done at a print shop. From 1930 until 1937, the effects of the Depression are apparent; programs were either typed or mimeographed and had no cover. The programs for two years are now missing. In 1940, the covers reappeared, this time typed on construction paper. The print shop printing never came back.

The last program in the box dates from 1964. I don't know the last year Mrs. Florence Johnson attended the club. I do know that Mrs. Johnson and her husband Turk celebrated their fiftieth wedding anniversary, for in the same box are some beautiful white fiftieth anniversary napkins with golden bells and the names Florence and Turk between the years "1920" and "1970." A neighbor told me that Mrs. Johnson died in 1981. The high school had closed in 1967. The lumber yard and hardware store closed about the same time but no one knows when for sure. The last gas station went after that.

Back to those programs. The Motto never changed. The sweet pea kept its standing. So did the pink and white club colors. The club collect which follows persisted month after month, year after year:

A COLLECT FOR CLUB WOMEN

Keep us, O God, from pettiness;
Let us be large in thought, in word,
in deed.

Let us be done with fault-finding
and leave off self-seeking.

May we put away all pretense and
meet each other face to face, without
self-pity and without prejudice.

May we never be hasty in judgment
and always generous.

Let us take time for all things;
make us grow calm, serene, gentle.

Teach us to put into action our
better impulses; straightforward
and unafraid.

> Grant that we may realize it is
> the little things that create differences;
> that in the big things of life
> we are as one.
>   And may we strive to touch and
> to know the great common woman's
> heart of us all, and oh, Lord God,
> let us not forget to be kind.
>
>   Mary Stewart

By modern standards, these people were poor. There was a kind of naïveté among these relatively unschooled women. Some of their poetry was not good. Some of their ideas about the way the world works seem silly. Some of their club programs don't sound very interesting. Some sound tedious. But their monthly agendas were filled with decency, with efforts to learn about everything from the birds to our government, and with coping with their problems, the weather, diseases. Here is the irony: they were living up to a far broader spectrum of their potential than most of us do today!

I am not suggesting that we go back to 1923 or even to 1964. But I will say that those people in that particular generation, in places like Matfield Green, were farther along in the necessary journey to become native to their places, even as they were losing ground, than we are.

Why was their way of life so vulnerable to the industrial economy? What can we do to protect such attempts to be good and decent, to live out modest lives responsibly? I don't know. This is the discussion we need to have, for it is particularly problematic. Even most intellectuals who have come out of such places as Matfield Green have not felt that their early lives prepared them adequately for the "official" culture.

I want to quote from two writers. The first is Paul Gruchow, who grew up on a farm in southern Minnesota:

I was born at mid-century. My parents, who were poor and rural, had never amounted to anything, and never would, and never expected to. They were rather glad for the inconsequence of their lives. They got up with the sun and retired with it. Their routines were dictated by the seasons. In summer they tended; in fall they harvested; in winter they repaired; in spring they planted. It had always been so; so it would always be.

The farmstead we occupied was on a hilltop overlooking a marshy river bottom that stretched from horizon to horizon. It was half a mile from any road and an eternity from any connection with the rest of the culture. There were no books there; there was no music; there was no television; for a long time, no telephone. Only on the rarest of occasions—a time or two a year—was there a social visitor other than the pastor. There was no conversation in that house.[2]

Similarly, Wallace Stegner, the great historian and novelist, confesses to his feeling of inadequacy in coming from a small prairie town in the Cypress Hills of Saskatchewan. In *Wolf Willow* he writes:

Once, in a self-pitying frame of mind, I was comparing my background with that of an English novelist friend. Where he had been brought up in London, taken from the age of four onward to the Tate and the National Gallery, sent traveling on the Continent in every school holiday, taught French and German and Italian, given access to bookstores, libraries, and British Museums, made familiar from infancy on with the conversation of the eloquent and the great, I had grown up in this dung-heeled sagebrush town on the disappearing edge of nowhere, utterly without painting, without sculpture, without architecture, almost without music or theater, without conversation or languages or travel or stimulating instruction, without libraries or museums or bookstores, almost without books. I was charged with getting in a single lifetime, from scratch, what some people inherit as naturally as they breathe air.

How, I asked this Englishman, could anyone from so deprived a background ever catch up? How was one expected to compete, as a cultivated man, with people like himself? He looked at me and said dryly, "Perhaps you got something else in place of all that."

He meant, I suppose that there are certain advantages to growing up a sensuous little savage, and to tell the truth I am not sure I would trade my childhood of freedom and the outdoors and the senses for a childhood of being led by the hand past all the Turners in the National Gallery. And also, he may have meant that anyone starting from deprivation is spared getting bored. You may not get a good start, but you may get up a considerable head of steam.

Countless writers and artists have been vulnerable to the "official" culture, as vulnerable as the people of Matfield Green. Stegner comments:

I am reminded of Willa Cather, that bright girl from Nebraska, memorizing long passages from the *Aeneid* and spurning the dust of Red Cloud and Lincoln with her culture-bound feet. She tried, and her education encouraged her, to be a good European. Nevertheless she was a first-rate novelist only when she dealt with what she knew from Red Cloud and the things she had "in place of all that." Nebraska was what she was born to write; the rest of it was got up. Eventually, when education had won and nurture had conquered nature and she had recognized Red Cloud as a vulgar little hold, she embraced the foreign tradition totally and ended by being neither quite a good American nor quite a true European nor quite a whole artist.[3]

It seems that we still blunt our sensitivities about our local places by the likes of learning long passages from the *Aeneid* while wanting to shake from us the dust of Red Cloud or Matfield Green. The extractive economy cares for neither Virgil nor Mary Stewart. It lures just about all of us to its shopping centers on the edges of major cities. And yet, for us, the *Aeneid* is as essential to becoming native to the Matfield Greens as the bow and arrow were to the paleolithic Asians who walked here across the Bering land bridge of the Pleistocene.

Our task is to build cultural fortresses to protect our emerging nativeness. They must be strong enough to hold at bay the powers of consumerism, the powers of greed and envy and pride. One of the most effective ways for this to come about would be for our universities to assume the awesome responsibility to both validate and educate those who want to be homecomers—not necessarily to go home but to go someplace and dig in and begin the long search and experiment to become native.

It will be a struggle, but a worthy one. The homecomer will not learn the likes of Virgil to adorn his talk, to show off, but will study Virgil for insight, for utility, as well as for pleasure.

We can then hope for a resurrection of the likes of Mrs. Florence Johnson and her women friends who took their collect seriously. Unless we can validate and promote the sort of "cultural information in the making" that the New Century Club featured, we are doomed. An entire club program devoted to coping with the heat of August is being native to a place. That club was more than a support group; it was cultural information in the making, keyed to place. The alternative, one might suggest, is mere air conditioning—not only yielding greenhouse gases but contributing to global warming and the ozone hole as well, and, if powered with nuclear power, to future Chernobyls.

Becoming native to this place means that the creatures we bring with us—our domesticated creatures—must become native, too. Long ago they were removed from the original relationships they had with their ecosystems and pressed into our service. Our interdependency has now become so complete that, if proprietorship is the subject, we must acknowledge that in some respects they own us. We humans honor knowledge of where we came from, counting that as baseline information, essential to our journey toward nativeness. Similarly we must acknowledge that our domesticated creatures are descendants of wild things that were shaped in an ecological context *not* of our making when we found them. The fence we built to keep their relatively tame and curious wild ancestors out of our Early Neolithic gardens eventually became the barbed wire that would contain them. At the moment of first containment, those fences must have enlarged our idea of property and property lines. When we brought that notion of property lines with us to this distant and magnificent continent, it was a short step to the invisible grid that in turn created the tens of thousands of hard and alien lines that dominate our thoughts today. Did the natives at Lone Tree Massacre foresee this? Those lines will be with us forever, probably. But we can soften them. We'll have to, for the hardness of those lines is proportional to our sense of the extent to which we own what we use. Our becoming native will depend on our emerging consciousness of how we are to use the gifts of the creation. We must think in terms of different relationships. Perhaps we *will* come to think of the chicken as fundamentally a jungle fowl. The hog will once again be regarded as a descendant of a roaming and rooting forest animal. Bovines will be seen as savanna grazers.

An extractive economic system to a large degree is a derivative of our perceptions and values. But it also controls our behavior. We have to loosen its hard grip on us, finger by finger. I am hopeful that a new economic system can emerge from the homecomer's effort—as a derivative of right livelihood rather than of purposeful design. It will result from our becoming better ecological accountants at the community level. If we must as a future necessity recycle essentially all materials and run on sunlight, then our future will depend on accounting as the most important and interesting discipline. Because accountants are students of boundaries, we are talking about educating a generation of students who will know how to set up the books for their ecological community accounting, to use three-dimensional spreadsheets. But classroom work alone won't do. They will need a lifetime of field experience besides, and the sacrifices they must make, by our modern standards, will be huge. They won't be regarded as heroic, at least not in the short run.

Nevertheless, that will be their real work. Despite the daily decency of the women in the Matfield Greens, decency could not stand up against the economic imperialism that swiftly and ruthlessly plowed them and their communities under.

The agenda of our homecoming majors is already beyond comprehensive vision. They will have to be prepared to think about such problems as balances between efficiency and sufficiency. This will require informed judgment across our entire great ecological mosaic. These graduates in homecoming will be unable to hide in bureaucratic niches within the major program initiatives of public policy that big government likes to sponsor. Those grand solutions are inherently anti-native because they are unable to vary across the varied mosaic of our ecosystems, from the cold deserts of eastern Washington to the deciduous forests of the East, with the Nebraska prairie in between. The need is for each community to be coherent. Knowing this, we must offer our homecomers the most rigorous curriculum and the best possible faculty, the most demanding faculty of all time.

Professor J. Stan Rowe describes what we might see by imagining that we humans could make ourselves small enough to enter some average-sized cell and, once there, continue to miniaturize to the point that we need binoculars to examine various parts off in the distance.[4] The parts with dynamic processes we would undoubtedly designate as living. There would be streaming cytoplasm, replicating DNA, and amino acids being hitched to one another in precise order, forming polypep-

tide chains of varying lengths. These along with the mito-
chondria, the cells' powerhouses we, with our human minds,
would likely designate as living, too. But if we have lost our
perspective during miniaturization we would inevitably designate some things—crystals, some membranes—as dead. If we
used this living versus dead taxonomy, it would only be because
we had, during our shrinkage, lost our ability to see the larger
perspective. Lacking our former, more comprehensive mind, we
would think that some things count more than others.

Now imagine that a proportionally large creature were to
arrive in our solar system and, after some shopping, pick the
Earth to visit. Imagine such a creature able to shrink to our
size but, unlike us, keeping the larger perspective in mind. He
or she or it would soon discover that a most amazing species
had changed the face of the earth in some dramatic ways to
grow food and fiber. Our visitor, at first glance, might think
that humans are basically artists bent on dressing the earth
with their own designs.

Now imagine that to our visitor the Earth is a sort of
field site, an object of study for something like a doctoral dis-
sertation. Maybe it is a long-term study in which the investi-
gator visits our planet every two hundred years or so beginning ten thousand years ago. On this, the fiftieth visit, this
student of the earth would realize that the population of one
particular species, self-named *Homo sapiens*, is rapidly increasing in number and that as it does so its members pollute and
destroy more and more of the parts of the planet necessary for
the maintenance of what they call life.

Many humans in thousands of places over the globe *have* been truly artful. But, as our visitor would realize, there is cause for alarm. During the two hundred years between the last and the present visit, many deposits of energy-rich carbon have been discovered in pools and seams and employed to power the human enterprise.

From the outside perspective, when the traditional cultures relied exclusively on contemporary sun power, the agricultural/cultural activity around and below could best be characterized as a form of intimacy. This activity was at once artistic and cultural. The body and the mind with the eye worked as one. There was no separation. Food was not just fuel then, and the tools necessary to capture sunlight and to provide food—air, water, soil—if they were dead, were dead in the same way as a crystal or a membrane is dead in the cell. This is no mere detail. The designation non-living invites a prejudice. Since air and water and soil are just dead stuff lying around, we act as though we can pollute or destroy them at will.

For the artistic farmer, the tool is not the brush but rather the pitchfork, the hoe, the rake, the shovel, the pruning shears, a team of horses, even a diesel tractor if it is run on vegetable oil. Whether in China, Peru, Africa, Sicily, or among the Hopi of the Southwest, the agricultural artist prefers wisdom over cleverness. He or she experiments, but experimentation is subordinate to tradition. The true artist honors balance of emotion and technique, people and land, individual and community, plant and animal.

Since the 1930s industrialized agriculture has been increasingly promoted by the industrial mind. But now a small but growing minority realizes that high energy destroys information of both the cultural and the biological varieties. This approach not only pollutes landscapes, it rooster-tails the finite supply of nutrients from our agricultural lands into the supermarkets, into the kitchen sinks, onto the chopping boards, onto the tables, and into the human gut, and, once there, more or less heads only one way, downstream into the sewers and graveyards.

Maybe our problem is that we are unable to keep the perspective of that outside observer, for we fail to absorb what we know: that nutrient cycles must be closed; that if we introduce into the environment chemicals with which we have not evolved, we must regard them as guilty until proven innocent; that fossil fuels are finite; that agriculture—not agri-business is the source of a heathy culture; that all parts are important to the whole and to other parts.

The realities of industrialization are all around us. No "ain't-it-awful" checklist is necessary. What we must think about, therefore, is an agriculture with a human face. We must give standing to the new pioneers, the homecomers bent on the most important work for the next century—a massive salvage operation to save the vulnerable but necessary pieces of nature and culture and to keep the good and artful examples before us. It is time for a new breed of artists to enter front and center, for the point of art, after all, is to connect. This is the homecomer I have in mind: the scientist, the accountant who converses with nature, a true artist devoted to the building of agriculture and culture to match the scenery presented to those first European eyes.

6 Developing the Courage of Our Convictions 🐾

Most of our modern assumptions are so deeply rooted that either we count them as "just natural" or we have no recognition as to what they really are. A major part of that consciousness comes from being raised in a society dominated by science and its technological arrangements, most of which would not be here without the high energy that comes from fossil fuel and nuclear power. We have a "high-energy consciousness," a monetarily cheap energy consciousness that is a mere blip in human history, but a consciousness that now "comes with the milk." (George Bernard Shaw once said that "perfect memory is perfect forgetfulness.") Even when we try to think about other possibilities, other worldviews, the powerful assumptions stirring within us reassert themselves in unexpected and often undetected ways. So our modern thinking is itself resistant to critique and change, even as the end of the fossil fuel epoch comes in sight.

At some level, most of us want to live within our means, to become native to this place at the very time the target appears to be receding faster than the shot we aim at it. We can accept the fact that where we were as a people in 1900 is different from where we were in 1850. But the difference between where we are in the 1990s and where we were in 1950 is vastly different and far greater. Thus our definition of "becoming native" is more seriously compromised than ever.

We have, though, learned some very important lessons during this high-energy phase of our journey. First off, in this world now dominated by economic thought, we have discovered that comfort and security are not solutions to the human condition and that affluence has not solved the economic problem. In fact, economic anxiety has increased and preoccupation with economic issues is higher than ever. Furthermore, economic development has led to enormous ecological destruction. But even if development were not a problem in terms of our relationship with the planet, even if it did not deplete the mines and the wellheads or poison the earth and deplete its atmosphere, even if there were an infinite supply of resources or infinite substitutability, development has been destructive of our relationships with people and with place.

With this in mind, as Harvard economist Steve Marglin says, we must go beyond the questions of income distribution, distorted accounting, and conspicuous consumption, and come to grips with the fundamental assumptions of economics: unlimited desire, absolute scarcity, calculation, and maximization.[1] During this era in which we had a conference in Rio where we talked about sustainable development—almost a contradiction in terms—we are now coming to understand another phrase, "cultural affirmation." As Marglin points out, because of the "paradox of affluence," which has sharpened the critique of economics, cultural affirmation adds a new dimension to the critique.

Cultural affirmation is on the line because cultural diversity is in decline. And now we see, we hope not too late, that the economic forces that destroy rainforest also destroy culture. As Mary Catherine Bateson said more than ten years ago, we face an "information crisis,"[2] no trivial matter, for cultural information like biological information is hard won. Famine, disease, death, pain—all negatives—inform the "do's" and "don'ts," probably more than the positive side of living. When we think of biotic diversity, those unique DNA arrangements of various species may be equal in the eyes of God but not to the forces of natural selection. Because of us, the selection pressures now wipe out much of that hard-won information, won in a sunpowered world. What characterizes the high-energy epoch, with the language of modern economics for justification, is a way of being that is both simple and simplifying. High energy *does* seem to destroy information of both the cultural and the biological varieties. It is a little bit like passing through a juvenile stage, whether of an ecosystem or of a teenager. There is an excess of potential energy but an inefficiency in the utilization of that energy. The sort of economic language that informs is based on the assumption of infinite resources or infinite substitutability. The parallel is sharp, for this is the operating assumption of the young who have been overly indulged. From childhood to young adulthood, responsible parents are obligated to teach their offspring important spring important limits they are likely to face as responsible adults. If the children are simply provided for and not exposed to or taught the reality of economic limits, as adults they are likely to accept a major assumption of our modern economists. It is easy to understand why, when we Europeans came to this abundant continent, we quickly adopted the assumptions of the spoiled youth. The spirit of this assumption was assisted by the emerging spirit of invention made possible with access to the huge supply of fossil fuels. But that high-energy era, let us hope, will end along with the era of cheap fossil fuels.

In the foregoing pages, I have tried to outline what is necessary to adopt an ecological worldview and find a better way while we still have slack. This won't be the first time we have experienced a major shift in consciousness. The ecological stakes were not so high when the churchmen approached Galileo and asked him to recant everything he had said and written about a heliocentric universe, but neither were they trivial. For people in that time, more was on the line than the physical structure of the universe; it was the structure to which the ancients had attached their moral code. If the physical cosmology was wrong, what about the moral code to which it was attached? Dante's notion of Heaven, Hell, and Purgatory (which was more or less isomorphic with the Church's notion) was not much amended in the minds of those who followed this heretic, Galileo, at least not at first. All those whom Dante met on his travels through the underworld were conveniently parked in circles going from bad to worse. The physical structure of the Inferno was not the subject of Galileo. It was the passage to Paradise through the planets upward to the Angelic Hierarchies, toward God himself, that required a redrawing of the map, for before Copernicus and Galileo, the agreed-upon journey from earth had a stopover first at the moon, then, in order, Mercury, Venus, the Sun, Mars, Jupiter, Saturn, and onward and outward. (They missed some, but that is a mere detail.) After Galileo, the earth was no longer at the center and the sun was no longer in the fourth position. The physical hierarchy for sins could stay intact but the physical hierarchy for virtue was shaken. Think of what that must have meant for the literal minded. In a similar manner, the structure that has assisted virtue in our time is now shaken. That structure is community. Too much was on the line for those churchmen to take a look through Galileo's telescope, but no more than when communities were being destroyed in our time. It is easy to understand why one of Galileo's old friends from younger days, a man who became pope, would force him to recant and then put him under house arrest. A different physical cosmology would lead to a shift in the religious code. And it did!

But the story is complex. In our time, we cite the early bravery and intellectual honesty of Galileo and are saddened by his persecution and withdrawal from his public position. We tend to see the churchmen as the bad guys and the church is further criticized for keeping his writing on the forbidden list for two hundred years. But what of our own dominant establishment—the scientific-technological establishment— insisting on the rightness of its methodology as the way to know and the products it spawns as essential to happiness? This establishment has a near worship of the idea of the separation of church and state, probably a good thing. But could we have a national debate about the need to separate science and state? The science-state alignment has been thousands of times more ecologically destructive than the church-state alliance ever was. For any of us who would object to such a debate, scientists or not, it seems fair to ask how they are different from the churchmen of Galileo's time.

Maybe the problem before us has to do with the tiresome job of learning to see what is before us and what the possibilities are. Oliver Sachs, a neurophysiologist and author of *The Man Who Mistook His Wife for a Hat,* has an article in the *New Yorker* that describes Virgil, blind from age four until his sight was restored at fifty following an operation. Virgil had to learn to see, but it was tiring work. (Development of that part of the cerebral cortex is dependent on being able to see along the way.) In a restaurant with his eyes open, Virgil could spear the vegetables in the salad early in the meal with accuracy, but he soon tired of locating and spearing them and finished his salad by eating with his fingers. A bowl with various kinds of fruit was set before him. Without looking, he would hold each piece in his hands, weigh it thoughtfully while his sensitive fingers felt the shape and surface, and then accurately name each one. A wax pear was slipped in and immediately he laughingly declared that it was a candle shaped like a pear or a bell. His never-blind companions could not make this fine distinction from feel. Eventually, Virgil lost his sight again, and one gets the idea that he was happy to be back in his old familiar world of darkness and touch.

Elements of this story make it a useful metaphor as we

think of countless scientists and technologists of our time. The scientific-technological revolution has given us the ability to "feel" subtle differences, that is, to know part of nature in depth. Because of sensitive instrumentation, we know about atoms now in a way we never knew before. We know about DNA, RNA, the code and protein synthesis, and that the code is universal. We know more about how old various rock formations are, what creatures were alive when they were laid down. We know about the cosmos. Who would deny that we have received countless benefits from this era?

Through science and technology we have probed and penetrated for utility and economic advantage. Like Columbus and Coronado who came to and penetrated this continent in search of gold, we have penetrated with our science, our "search for gold." Perhaps that is a bit too strong. We cannot imagine what the scale of destruction would have been had the mental and physical penetration been limited to the desire to "know how the world works." But motivations are seldom pure, or if they are initially, seldom do they stay that way. What if we had decided, for example, to learn about the nature of health? We take health for granted, so we have given it little value. It is precisely because disease pays that we have little understanding of the nature of health. The "conquest" of disease has become a modern search for gold, especially for much of the current medical establishment. Much of the motivation for high yield in agriculture was in the "search for gold" realm and clearly is predicated on a form of conquest. High yields still have standing in people's minds over the ecological costs that industrialized agriculture accelerated.

Forget the motivation for a moment, whether the scientific-technological revolution was for pure knowledge or pecuniary goals. Metaphorically speaking, we may have the sensitivity to know the difference between wax fruit and real fruit from feel, but at the expense of our ability to experience the whole. And so here is a worry. Virgil seemed glad to retreat to his former world, even though he had added very briefly a necessary missing faculty for wholeness, because for forty-six years his brain had not been trained by the light it required. Because Virgil lacked en-*light*-enment, he preferred his deficiency. Here are the troubling questions: Will we modern scientists also become frustrated trying to develop a whole view now that we have gone so long into a culture of reductionism? Will we so enjoy the pleasure that comes from the exactness of knowing the difference between the scientific analogs of wax fruit and real? Or, can we simply know and enjoy the facts accumulated during this scientific-technological revolution and count our past discoveries as part of our inventory of knowledge, like our knowledge of geography or history, and reject the methodology, not altogether but *as the dominant way of discovering*? Can we move it to a subordinate role? I think that the answer is "yes," based on our emerging research agenda at The Land Institute. But, just as it takes courage to force the medical establishment to focus on the context of health rather than the biotic agents of disease, so it takes courage for agricultural researchers to risk looking downward from the ecosphere and seeing nature's ecosystems in the mosaic as primary objects of study. If we can do that, then we can fashion a new research agenda for agriculture featuring a dialectical interaction with nature and, ultimately, a conversation with nature. Hope is emerging from various quarters. A few economists have come out of the closet to admit that many of their important assumptions are wrong or meaningless. They see how their models are about as irrelevant as the Ptolemaic model held by the cardinals in the Vatican after Galileo. More modern economists will have to admit that much of what is important to the life-support system and culture does not compute. If all the world increasingly becomes chattel, then all of the world is a mine rather than a source of hope.

This does not represent a call for a return to a former state. Quite the opposite. For if there is any lesson from what we understand about the nature of the universe, from the Big Bang to the present (or even if it turns out to be a pulsating universe or whatever), change is the rule. No species ever returns from extinction no matter how much we might long for that to happen. No culture ever returns from extinction either, and though we may give tribute to a time past by something like the restoration of the likeness of Colonial Williamsburg, Williamsburg is as extinct as the dinosaurs. The restoration is like plaster over the bones of a museum dinosaur.

I hope no one feels that this discussion is limited to the fields of agriculture, medicine, and economics, for it is the wholeness of community life and the need for community life that are on the line. So when we think about the revitalization of small towns and rural communities worldwide, rather than insisting that we go back, I am instead insisting that we be careful as we go forward to avoid several impulses. Some of us might be tempted to gentrify the small places, make them Ecotopias or EcoDisneylands and replace every piece of every picket fence that existed a hundred years ago. Ecological community accounting won't work that way. Rather, it requires an assessment back to the source of energy and materials and an estimate of what the community costs are to provide this or that building material or gadget. We have to analyze as many costs as possible when materials and energy are moved through a boundary of interest, be it a county, a village, or anything else.

This accounting on the material and energy side of things is only the most obvious way to begin to develop our standards for right livelihood, a first step toward removing ourselves from the extractive economy. But there are also our biological origins, the way we were made. We must keep before us the need to gain and maintain a realistic image of what it was like for creatures of the upper Paleolithic as they gathered and hunted as a cultural species in numerous tribal arrangements for two million years, 150,000 years or so with the big brain. There are some Paleolithic predispositions for which we are "hardwired." Give a lecture in a small to large room and have someone enter from the back or side and quietly come forward to find a seat. Attention is largely drawn away from the speaker for a moment. This late arriver may be a surrogate hyena who might be looking for an opportunity to snap up an unattended child or perhaps a surrogate warrior from another tribe. Or what of the breast-feeding mother who hears a child—it need not be her own—cry. The milk pours forth, totally unbidden. Countless other examples of Paleolithic responses are available. It is dangerous to go too far with this sort of thinking, of course, but it is also dangerous or stupid to ignore the origins of much of our social behavior.

On the negative side, much of the time in our Paleolithic past we could likely take from the world with little thought for the morrow. Nature provided. And in most of the past, any technological device we developed must have increased our adaptive value. We did not have to be psychologically predisposed to put the brakes on technology. If we were as repulsed by technological proliferation as we are by spoiled meat crawling with maggots in a hot sun, it would be a very different world. Because we have "stolen fire" or because we are "fallen," we are stuck with creating the cultural language to develop the resistance.

On the positive side, much of what we do unconsciously in community that had an adaptive value in tribal life is beautiful and adaptive still. Think of the members of a community who respond to a neighbor's fire or flood or other physical disaster or attend to the retarded or the sick.

Why community works, how certain social arrangements come about, is a complex mystery in the same sense as why the liver got to be the way it is, carrying out its numerous functions as part of an integrated body. Some people do study the liver and its connections, but most of us simply accept it, are glad that it works, and seek to keep it healthy by keeping our bodies healthy. Do we need to understand community as well as we understand the liver? In the main, all we have to do is *provide the context* for community to happen and live in ways that will *keep it healthy*. From there on, much good will naturally occur. Our main task, then, is to commit ourselves to being diligent enough to fend off any effort that might threaten community. This is no small matter for there are plenty of forces around and within that will destroy it. The world is full of people who will encourage others to do things that will damage their liver. Sniffing of certain glues will do it. So will too much alcohol, or working unprotected in certain chemical factories. In a similar manner, the forces of power, particularly corporate power, are impatient with what is adequate for a coherent community. Because power gains so little from community in the short run, it does not hesitate to destroy community for the long run. The malls at the edge of town are a perfect example. We forget why they were built. Their designers did not say, "Let's make them ugly, wasteful, and devoted to consumerism." They turned out a design to export wealth *away* to their stockholders, most of whom reside in distant cities. Malls are suction pipes, designed to export regional wealth.

Were Matthew Arnold alive today, he might still see us "wandering between two worlds, one dead, the other powerless to be born."[3] I would put it another way. The quantum physics people tell us that for any given particle, it is impossible to predict its position with any reasonable degree of certainty, but that aggregations of atoms will yield high predictability. For the historian it is the other way around; the position of historical fact can be highly certain but aggregations of historical facts usually remain as aggregations with no overriding, generalizable sense. Those are the worlds we stand between, and neither is dead nor powerless to be born. They are reality, and it is ecology that holds them both.

In the time span since Coronado ordered the first murder of a true native on what is now Kansas soil, we of Western civilization have moved from the church, to the nature-state, to economics as the primary organizing structure for our lives. We have been through the hypocrisy of the church, the atrocity of the nation-state that peaked with Hitler, and now we are devotees of economics, the encoded language of human behavior that directs us toward ecological bankruptcy. It is time to move more aggressively on to the fourth phase, already under way, ecology.

The skeptical have reason to doubt the validity and power of an ecological view still poorly developed in its implications, in its reach. In all segments of society, perhaps we are about as far along as we are in our efforts at The Land Institute to build an agriculture for grain production based on the principles of a natural ecosystem such as a prairie. We are at about the same stage as the Wright brothers were at Kitty Hawk. That first flight was only a few feet high and long. The skeptical and impatient mind can sometimes be a terrible combination. Imagine such a mind watching those efforts at Kitty Hawk, trying to anticipate such a craft as a future carrier of as many people as then rode a train. That first plane showed little promise. But what was being tested was not an airplane but a principle, a principle that would one day lead to the likes of the 747 and the

SST. (I happen to think that the airplane is a bad idea and that we are really not meant to fly, considering the horrendous resource cost.) Because the ecological paradigm for agriculture is at about the stage of the Wright brothers at Kitty Hawk, we need more experiments, more time in the wind tunnel. But given patience and time, surely we can develop an agriculture that will be here long after anything like the airplanes of today have grounded. But it cannot happen in isolation.

The penetration of the ecological paradigm—using nature as the measure—into all of society will have to inform our economics, our health, our communities. Specialists everywhere, from the artist to the scientist, will have to operate in the spirit in which the late Wallace Stegner wrote. Wendell Berry in a tribute said this of him: "There must have been a moment when he decided that he would not be the kind of writer who would look on his native country as 'raw material' for his art, and leave it otherwise to take care of itself or to be cared for by other people, but that he would be a kind of writer who would be devoted to his country for its own sake, and do what he could to protect it."[4] We have to become that sort of patriot.

🐚 Finally, I come back to 1541, to Coronado and those thirty or so avarice-driven adventurers who made the side trip "northward by the needle" from the plains of Texas toward the land of Quivira, where they had been told by their guide, a native Indian slave, that there was gold. When it was clear that there was no gold, Coronado allowed this native of the land that would become Kansas to be strangled with a rope twisted about a stick. What was his offense? He had told a series of lies to men made gullible by greed. He was no fool, and he must have known the risk, but he did it anyway and he did it for one reason. He was homesick. Because he was a slave, the lure of gold was his ticket home. He thought he could outwit them in the end, but he failed. He was not cunning enough to overcome the power of conquest. The homecomer of today still confronts that power.

Notes 🦫

Prologue

1. We can imagine a future in which the fossil fuels will be stretched out for a long time to come: at the end of oil, then natural gas, then liquid fuels from coal. Because of global warming, we should probably be more worried about the abundance of fossil fuels than their short supply.

1. *The Problem*

1. Edith Connelley Ross, "The Quivira Village," *Collections of the Kansas State Historical Society, 1926-1928* 17: 514-34.

2. *Visions and Assumptions*

1. Quoted in Carl O. Sauer, *The Early Spanish Main* (Berkeley: University of California Press, 1966), 69.

2. Dan Luten, "Empty Land, Full Land, Poor Folk, Rich Folk," *Yearbook of the Association of Pacific Coast Geographers* 31 (1969).

3. Wallace Stegner, *Wolf Willow* (New York: Penguin, 1962), 59.

4. Quoted in Carolyn Merchant, *The Death of Nature* (New York: Harper and Row, 1983), 168.

5. Richard Levins and Richard Lewontin, *The Dialectical Biologist* (Cambridge, Mass.: Harvard University Press, 1985), 152-60.

6. Stephanie Yanchinski, "Boom and Bust in the Bio Business," *New Republic* 22 (January 1987).

7. Julie Ann Miller, "Mammals Need Moms and Dads," *Bioscience* 37 (June 1987): 379-82.

3. Science and Nature

1. Douglas Sloan, "Imagination, Education and Our Postmodern Possibilities," *Revision* 15, no. 2 (Fall 1992): 42-53.

2. Wendell Berry, "Nature as Measure," in *What Are People For?* (Berkeley: North Point Press, 1990), 208-9.

4. Nature as Measure

1. Lynn White, Jr., "The Historical Roots of our Ecological Crisis," *Science* 155 (10 March 1967): 1203-7.

2. Leonardo Boff, *Saint Francis: A Model for Human Liberation* (New York: Crossroad Publishing, 1984), 97-100.

3. Ronald Coase, "The Problem of Social Cost," *Journal of Law and Economics* (1960), 1-44.

4. Jerome Ellig, "Nobel Prize for Common Sense," *Wichita Eagle-Beacon*, March 29, 1992, 14A.

5. Published as "A Practical Harmony" in *What Are People For?* 102-8.

6. Virgil, *The Georgics*, trans. Smith Palmer Bovie (Chicago: University of Chicago Press, 1966), 5.

7. Berry, "A Practical Harmony."

8. Liberty Hyde Bailey, *The Outlook to Nature* (New York: Macmillan, 1905).

9. Liberty Hyde Bailey, *The Holy Earth*, 1915 (rpt. Christian Rural Fellowship, 1946).

10. Sir Albert Howard, *An Agricultural Testament* (New York: Oxford University Press, 1943; rpt. Emmaus, Penn.: Rodale Press, 1976), 4.

11. J. Russell Smith, *Tree Crops* (New York: Devin Adair, 1953; rpt. Washington D.C.: Island Press, 1987), 11.

12. Personal communication.

13. H.T. Odum, *Environment, Power and Society* (New York: Wiley Interscience, 1971), Fig. 3.5, pp. 73-74.

5. Becoming Native to Our Places

1. The expression "unwittingly accessible" is from Carlos Castenada, *Tales of Power* (New York: Simon and Schuster, 1974).

2. Paul Gruchow, "America's Farm Failure," *Small Farmer's Journal* (Summer 1991).

3. Stegner, *Wolf Willow*, 24-25.

4. Stan Rowe, "Viewpoint," *Bioscience* 42, no. 6 (1992): 394.

6. Developing the Courage of Our Convictions

1. Oral communication, Amherst, Mass., 1992.

2. Oral communication.

3. Matthew Arnold, "Stanzas from the Grande Chartreuse" (1855), stanza 15.

4. Wendell Berry, remarks at a memorial service for Wallace Stegner, Stanford University, May 3, 1993; unpublished.